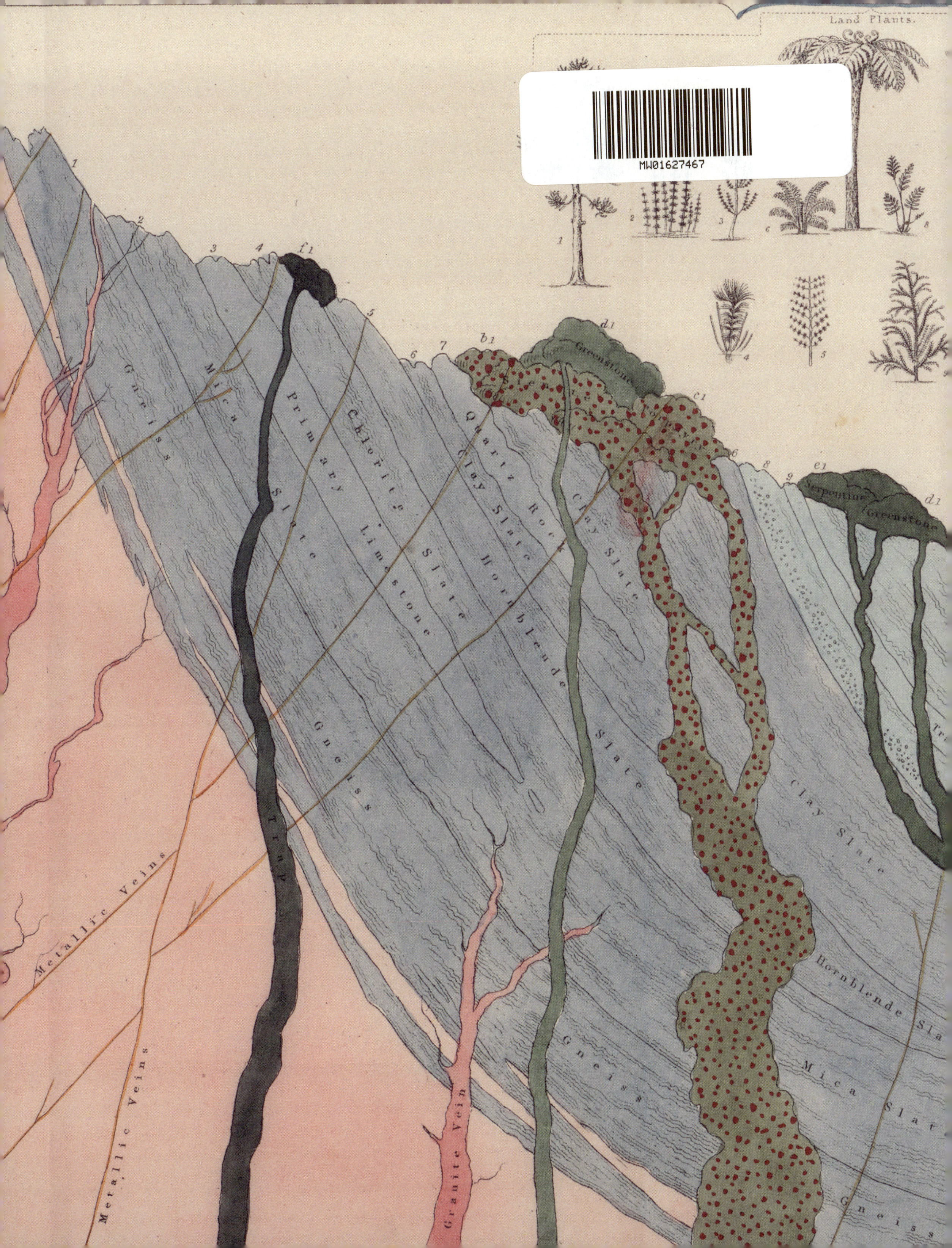

Land Plants
MW01627467
Gneiss
Mica
Primary Slate
Chlorite Slate
Limestone
Quartz Rock
Clay Slate
Hornblende
Greenstone
Serpentine
Clay Slate
Gneiss
Trap
Metallic Veins
Metallic Veins
Granite Vein
Gneiss
Gneiss

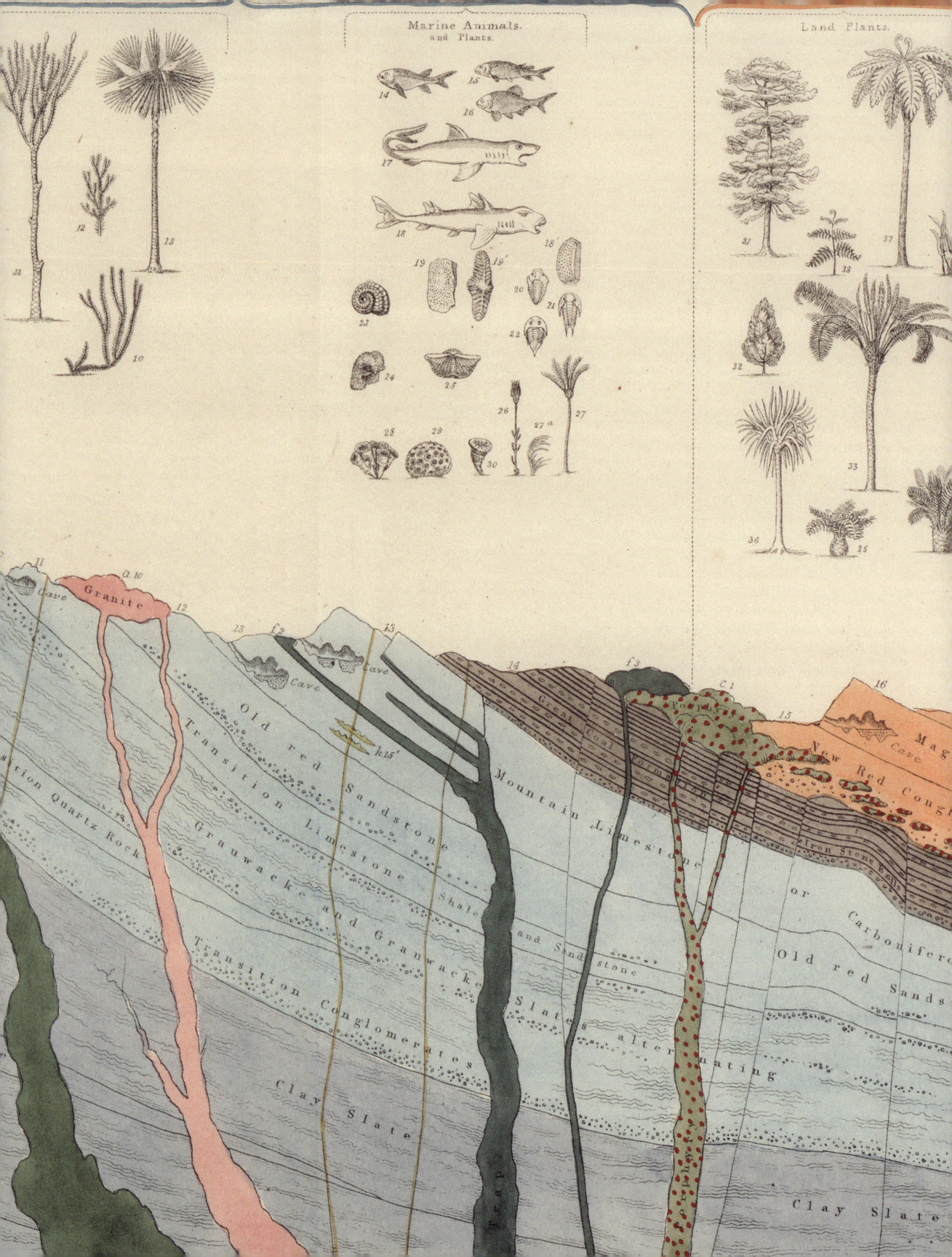

Marine Animals.
and Plants.
Land Plants.
Granite
Cave
Old red Sandstone
Transition Limestone
Grauwacke and Grauwacke
Transition Conglomerates
Clay Slate
Mountain Limestone
Shale and Sandstone
Slates alternating
Great Coal Formation
New Red Cong
Iron Stone Balls
or Carbonifero
Old red Sands
Clay Slate
Trap
Porphyry

Storm Cloud

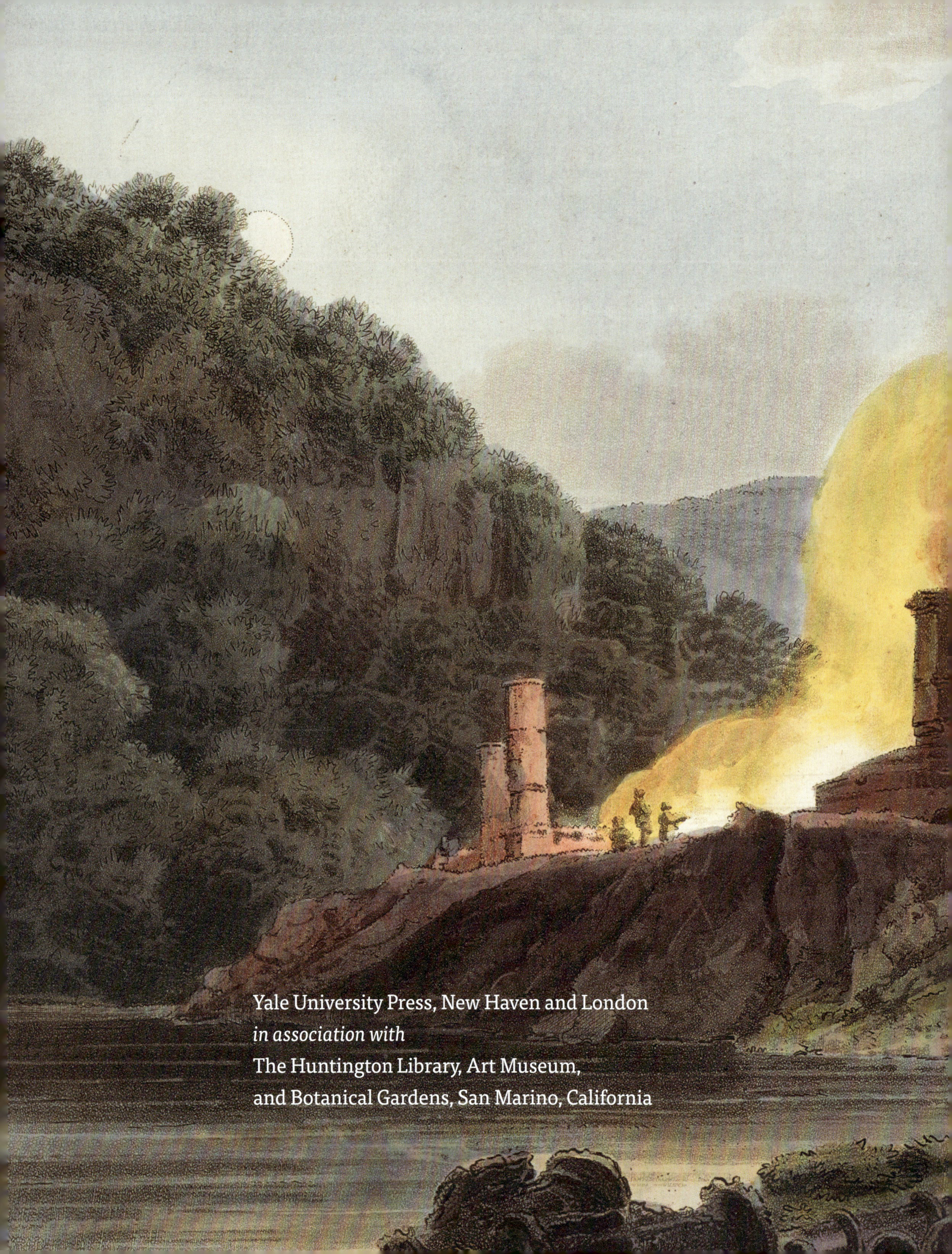

Yale University Press, New Haven and London
in association with
The Huntington Library, Art Museum,
and Botanical Gardens, San Marino, California

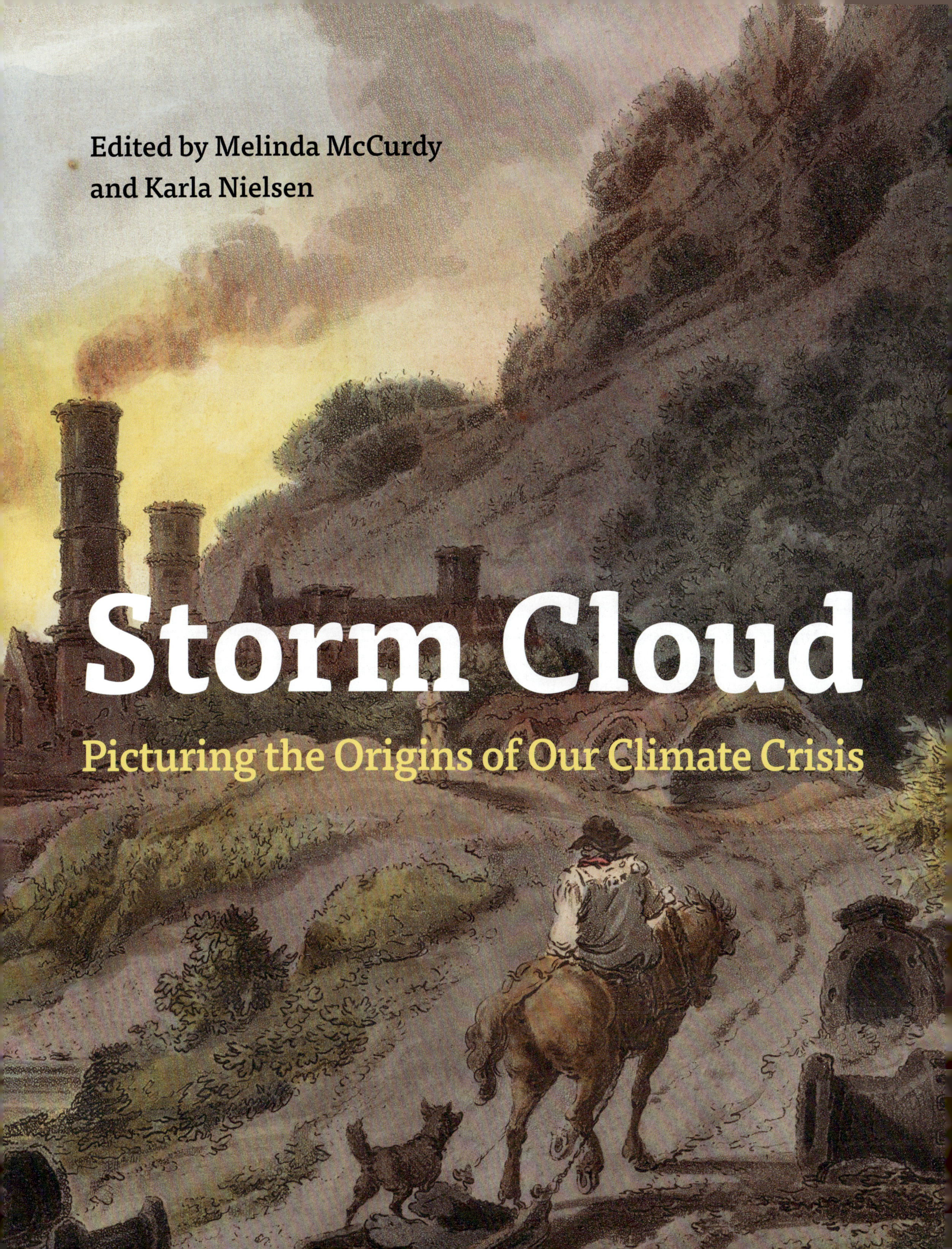

Edited by Melinda McCurdy and Karla Nielsen

Storm Cloud

Picturing the Origins of Our Climate Crisis

Contents

Directors' Foreword

Sandra Brooke Gordon and Christina Nielsen

Across the arc of the long nineteenth century, some five generations witnessed unparalleled revolutions in both industrialization and science. This book and the exhibition it accompanies trace the ways in which influential artists and writers in the United States and the United Kingdom reflected the changes to the natural world and the transformations in human perspective these revolutions set in motion. The authors remind us that current discussions about the influence of human activity on the environment—what John Ruskin ominously called the "storm-cloud"—can be traced back much further than the last few decades.

Here, at The Huntington, we are surrounded by bountiful nature and the artifacts of science and history. In the gardens we mark the passage of time in short bursts, say the bloom of a rose, or longer spans, such as a hundred-year-old Montezuma cypress. The books and manuscripts in our library, and artworks in our museum, are messages cast into the ocean of time hundreds or even thousands of years ago. Looking up at the nearby San Gabriel Mountains, we are constantly reminded of deep geological time counted in millions of years. And the visible domes and towers of Mount Wilson Observatory, where just a century ago the astronomer Edwin Hubble gathered evidence for an expanding universe, turn our minds to a calendar clocked in billions of years. Yet just a few years ago, we were sorely reminded of the fragility of the bounteous environment we take for granted when the Bobcat Fire burned ferociously for weeks across our beloved San Gabriels and produced a dolorous cloud of smoke and spread ash across the valley. When daytime turned dark, we were chastened to reflect on our own role as agents in global-scale events.

Part of The Huntington's response to the Getty-led initiative for 2024–25, PST ART: *Art & Science Collide*, this book and exhibition look to the literary and visual arts and scientific thinking on both sides of the Atlantic, to pose provocative questions of interlocutors, past and present. Looking at the past from multiple disciplinary perspectives, Huntington Library and Art Museum curators demonstrate how the coalescence of humanistic and scientific ways of knowing yields a deeper awareness of where we have come from, how we arrived at this juncture, and how we might build a better future together.

We have many people to thank for this exhibition, starting with the curators Melinda McCurdy and Karla Nielsen. They have made us see and think differently about the past and about the present. They, in turn, have been supported and inspired by the larger curatorial corps across The Huntington, in the Art, Botanical, and Library divisions.

President Karen Lawrence has unfailingly supported this project from its inception, and her rallying cry of "One Huntington" has allowed for greater cross-divisional work than ever before. As always, we are grateful to Senior VP and CFO Janet Alberti and Patty Hanna in Finance; Senior VP for Advancement and External Affairs Randy Shulman and Sarah Basile in Advancement; Holly Moore and the Conservation and Preservation department; Cynthia Tovar, Mark Jones, Sean Kennedy, and the Registration and Preparators team; Lana Johnson and the Exhibitions team; Director of Research Susan Juster; Jean Patterson in Publications; Thomas Polansky and the facilities staff and security officers; and Elee Wood and the team in Education and Public Programs, including our docents. We also extend profound thanks to Nicole Cavender, Director of the Botanical Gardens, who has encouraged us to focus on regeneration, and to consider meeting our needs today in ways that will not prevent future generations from meeting theirs.

Finally, we thank The Getty Foundation and other funders: the Tianqiao and Chrissy Chen Science Initiative, the Douglas and Eunice Erb Goodan Endowment, the National Endowment for the Arts, The Gladys Krieble Delmas Foundation, The Neilan Foundation, The Ahmanson Foundation Exhibition and Education Endowment, The Melvin R. Seiden-Janine Luke Exhibition Fund in memory of Robert F. Erburu, and the Boone Foundation.

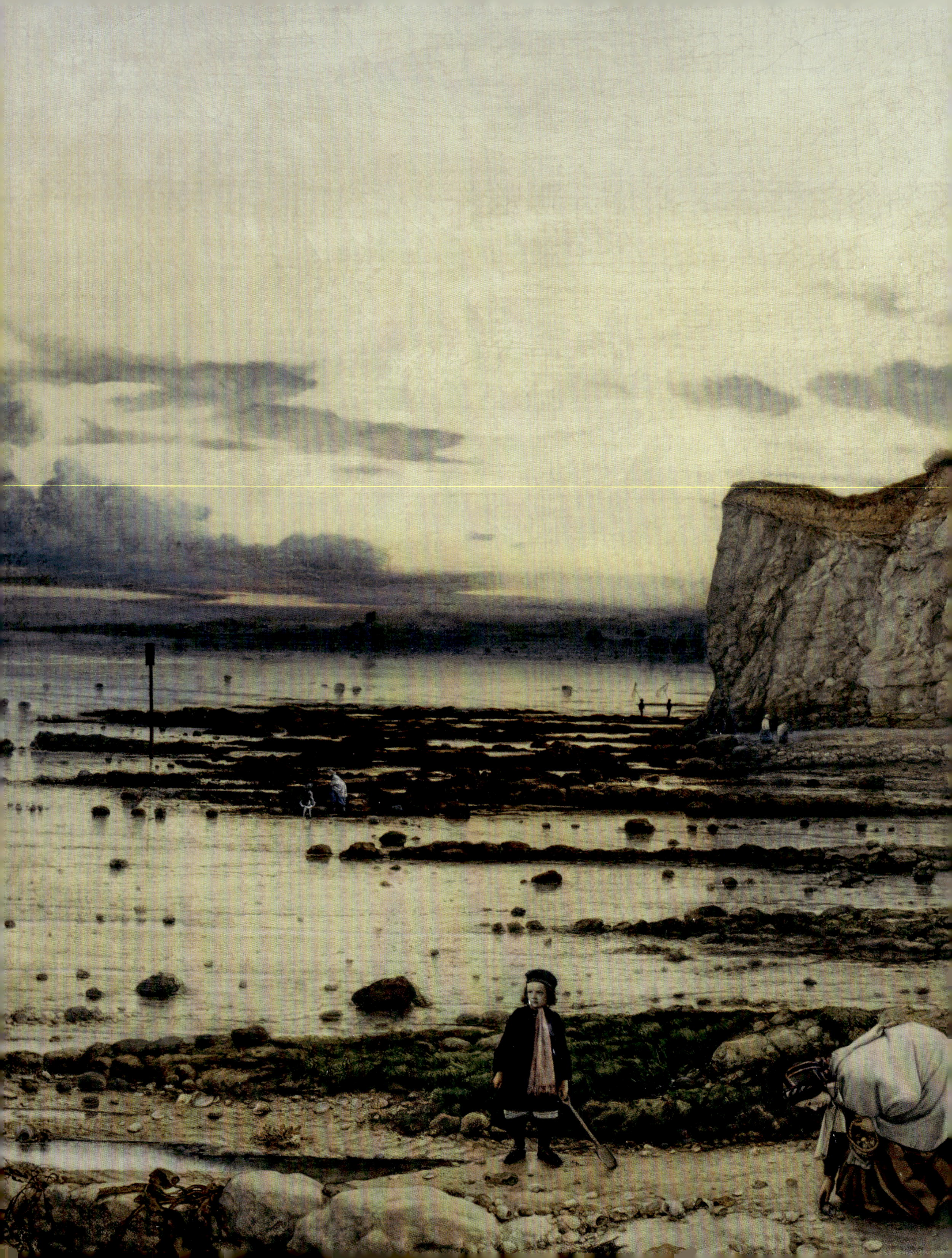

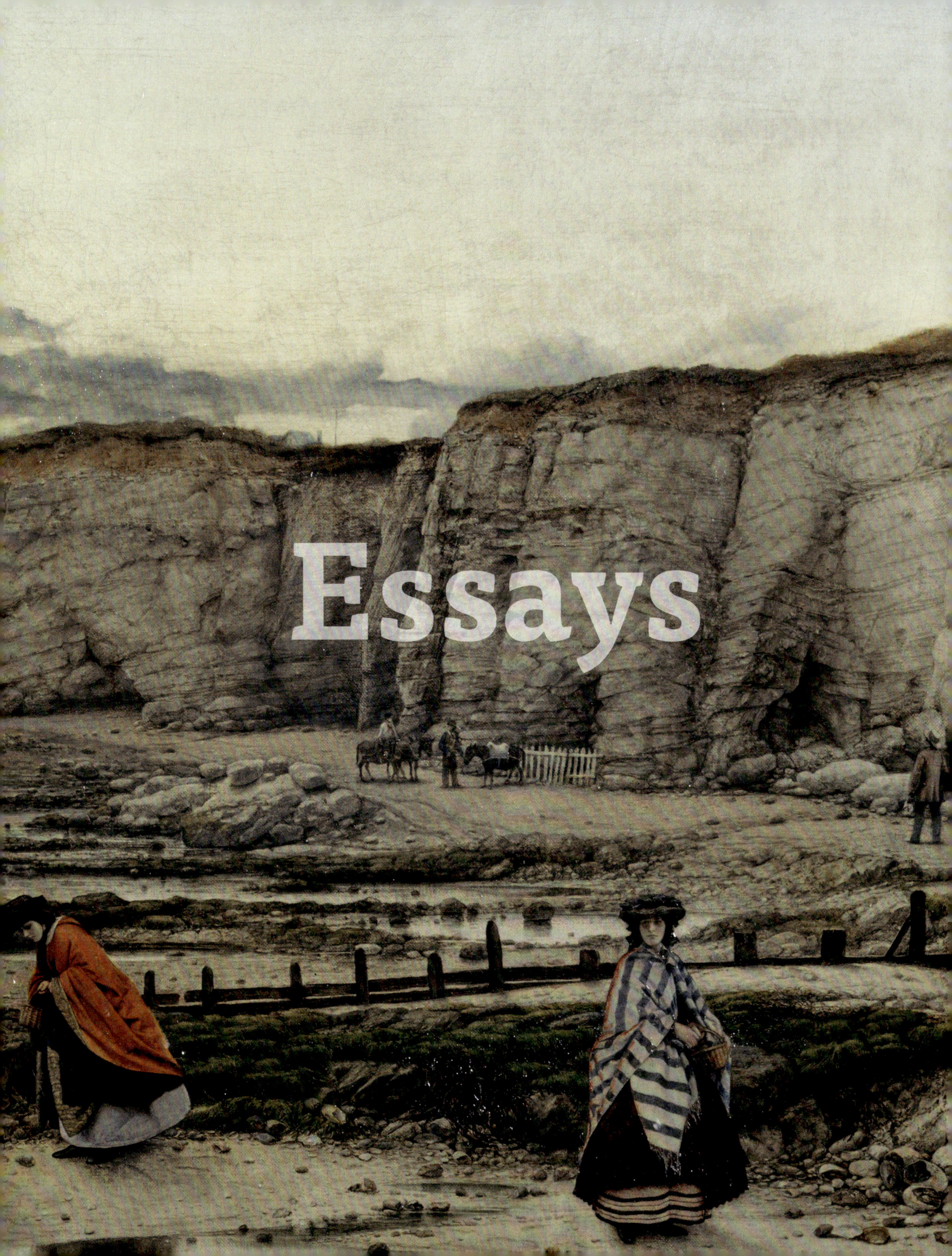

Essays

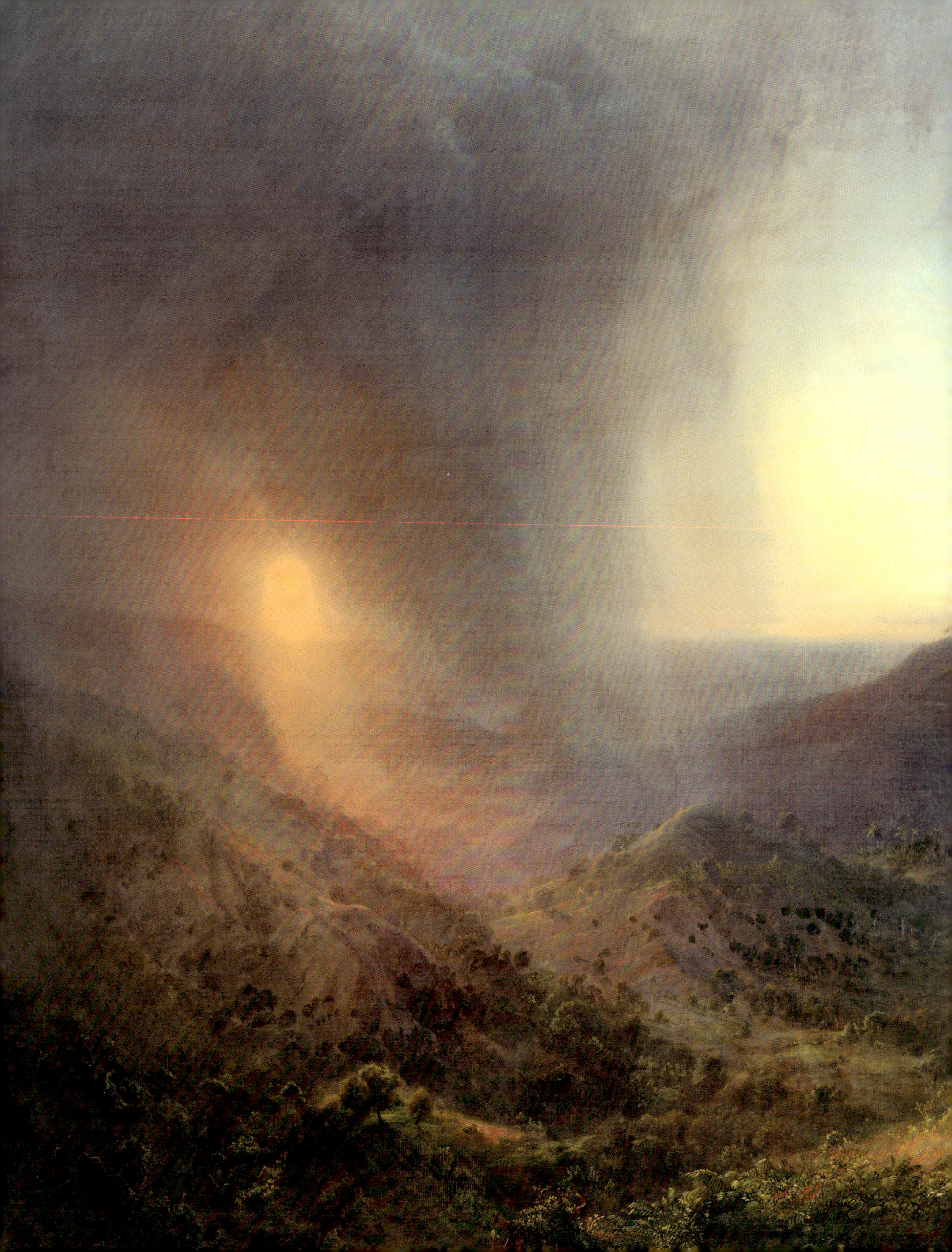

Melinda McCurdy and Karla Nielsen

Seeing through the Storm Cloud

Storm Cloud: Picturing the Origins of Our Climate Crisis traces a growing awareness of the industrial world's effect on the environment over the course of the long nineteenth century (roughly 1780–1930), as expressed through the period's art and literature, and in conversation with its evolving sciences. The title comes from a lecture series given by the Victorian writer and art critic John Ruskin in 1884. In "The Storm-Cloud of the Nineteenth Century," he described how decades of observing the English skies led him to conclude that his own age had created a disturbing environmental phenomenon. Ruskin's "storm-cloud" was smog, air pollution caused by carbon particulates pumped into the atmosphere by coal-burning homes and factories (fig. 1.1). Though he wouldn't have understood it in this way, his lectures became one of the earliest-published considerations of anthropogenic, or human-caused, climate change.[1]

As is now evident, the warming of our planet quickly escalated in the late twentieth century, the result of huge population growth, and increased use of fossil fuels—to run our cars, move our goods, and power our homes and industries—following the Second World War. The rapid increase of these and other earth-altering activities is known as the "Great Acceleration," and may be seen vividly in the graph of the Keeling Curve (fig. 1.2). The curve charts the measurement of carbon dioxide in Earth's atmosphere, using readings taken daily since 1958. Earlier numbers are calculated via ice core data and show how the 150 years surveyed in this exhibition set up the conditions for the upward trend. While not as dramatic as the surge of the last 70 years, the nineteenth century's nascent incline resulted from the establishment of many of the manufacturing, infrastructural, and consumer dynamics that caused the Great Acceleration. For Ruskin's age was one of rapid industrialization, years marked by innovation, invention, and expansion that remade the world and the place of humans on it.

Concurrent with the improvements in mining, farming, and machinery that spurred and sped up industrialization were the scientific advancements of

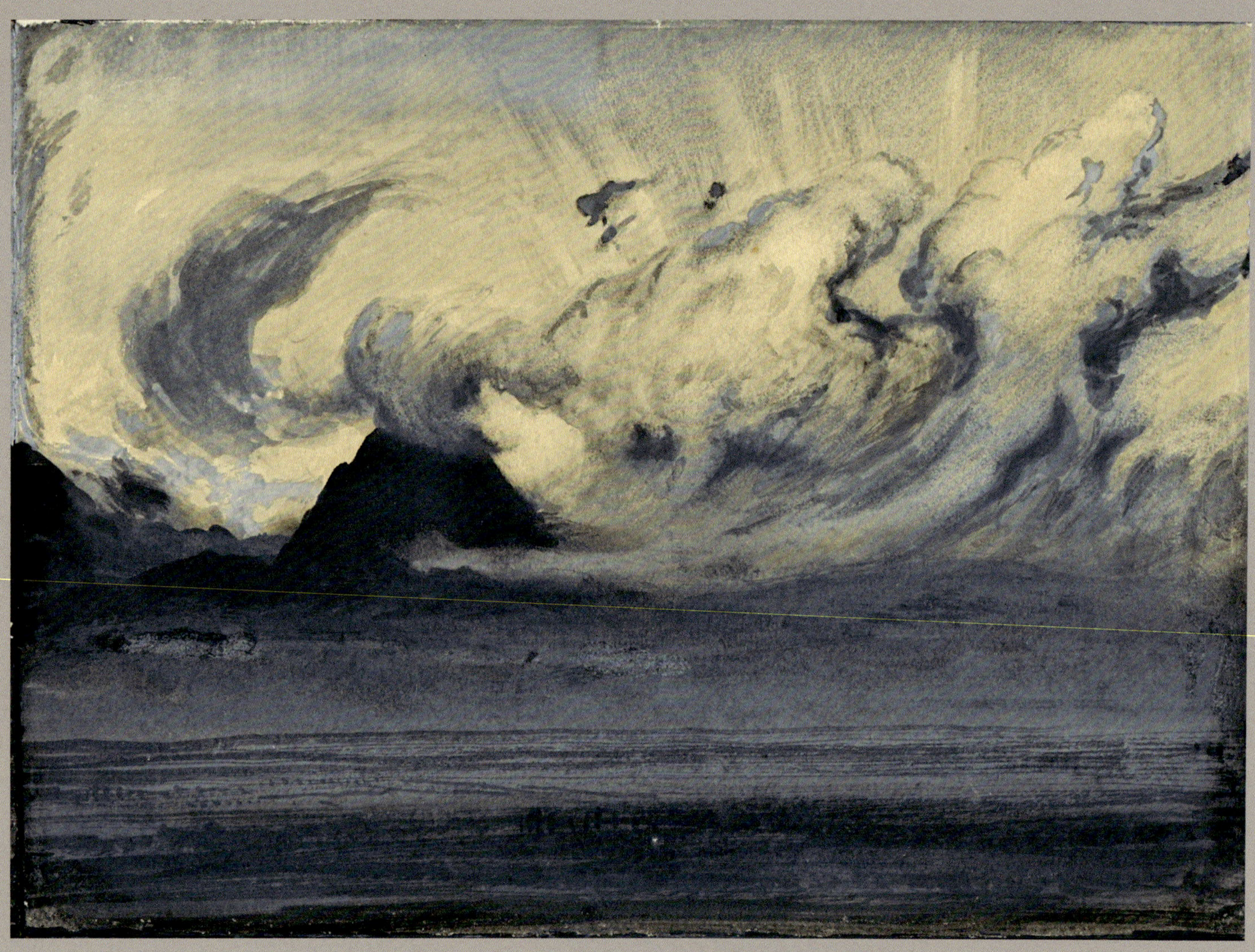

Fig. 1.1. Arthur Severn, after John Ruskin, *Thunderclouds, Val d'Aosta,* c. 1884 after 1858 original. Watercolor and gouache on paper, 5 × 7 in. (12.7 × 17.8 cm). The Ruskin, Lancaster University, 1996P1217.

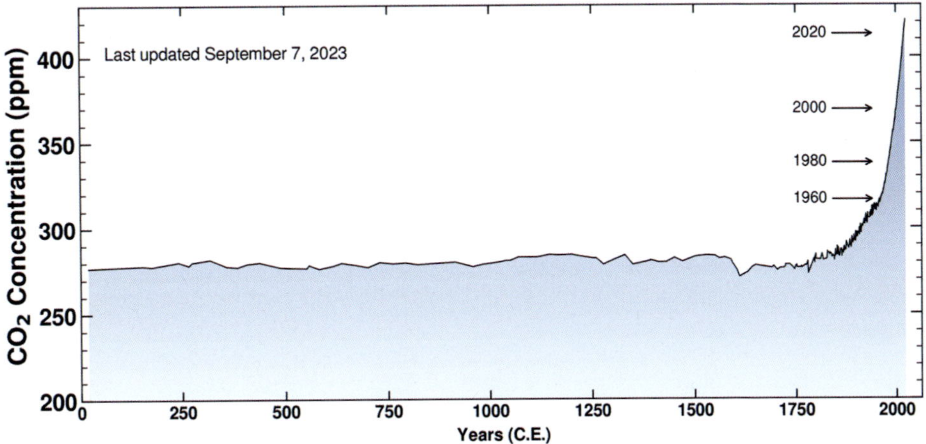

Fig. 1.2. Mauro Rubino, David Etheridge, David Thornton, Colin Allison, Roger Francey, Ray Langenfelds, Paul Steele, Cathy Trudinger, Darren Spencer, Mark Curran, Tas Van Ommen, and Andrew Smith, *The Keeling Curve* (2019): "Law Dome Ice Core 2000-Year CO_2, CH_4, N_2O and d13C-CO_2," 2019, CSIRO Data Access Portal, https://doi.org/10.25919/5bfe29ff807fb.

the nineteenth century through which we now understand industrialization's effects. Ruskin's was the era in which the sciences of geology, paleontology, and meteorology developed into the disciplines we recognize today, and in which the field of natural history began to encompass studies that would later be termed "ecology." Not long after the birth of the industrial revolution, voices from these sciences began to articulate a new understanding of the relationship between humans and the planet. In turn, artists and writers, such as Ruskin, began to engage with emerging scientific concepts, expressing in their own work a novel perception of humanity's place in, and impact on, the natural world. Because pinpointing a time and place for the advent of a new understanding of nature is impossible, and necessarily arbitrary, let us begin here: just before the last decade of the eighteenth century, in Hampshire, England.

Living with Nature

In 1789 English parson Gilbert White published *The Natural History and Antiquities of Selborne,* a study of the seasons, plants, and animals that "surround" the human inhabitants of a mostly rural county in the southeast of England. White described the flora and fauna of Selborne—what times of year they could be observed and where they could be found, and how they were connected in "the economy of nature."[2] The attention he paid to the interactions between life-forms, and his contention that all plants and animals depended on one another, makes him an early ecological thinker. His mode of viewing and recording came to define the conventions of natural history writing for many generations. Over the following two hundred plus years *Selborne* would go into more than four hundred editions and be translated into seven languages, exerting a huge influence on the study of the botanical and zoological natural world.[3] We know that Henry David Thoreau owned a copy of White's work. His book

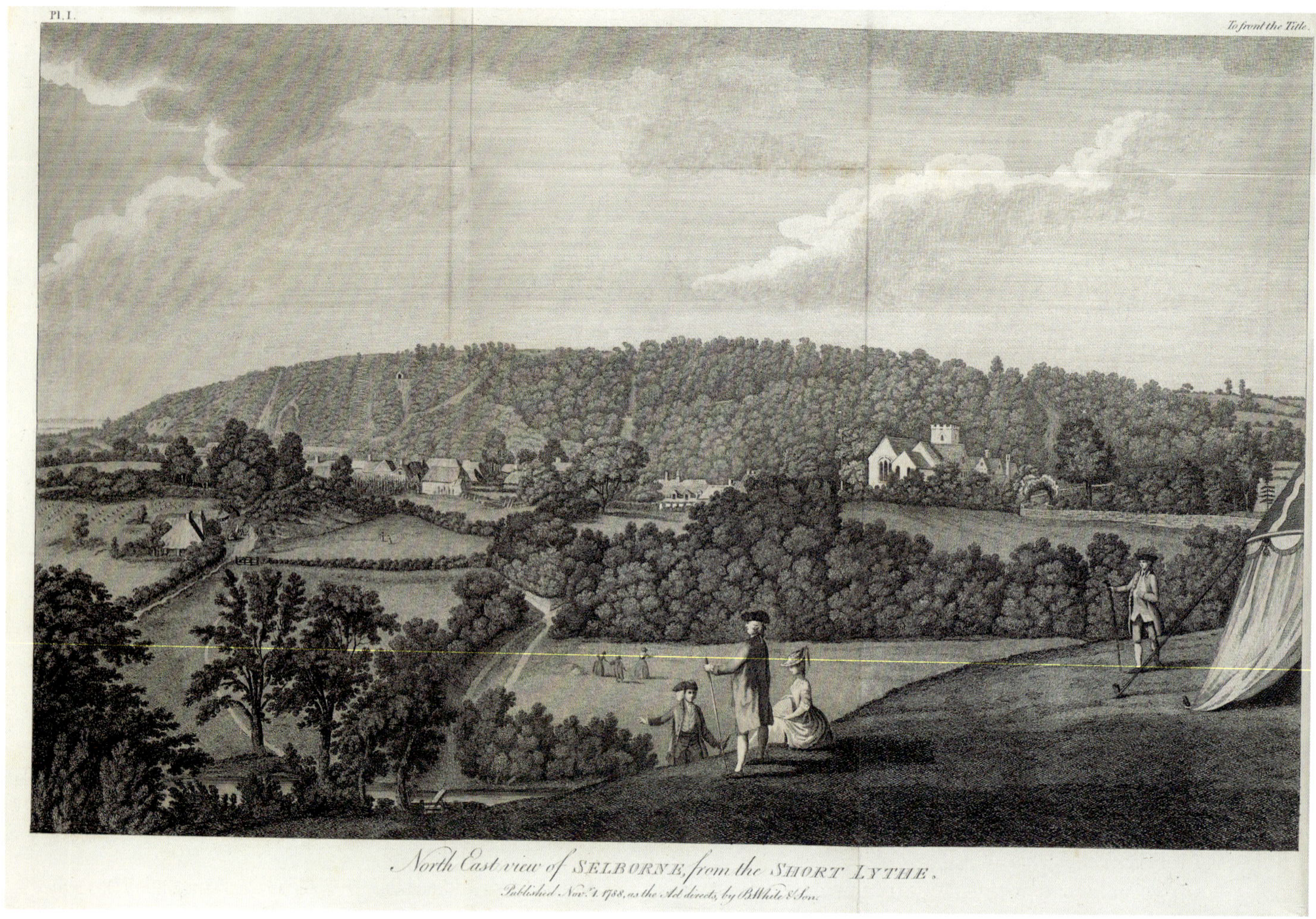

Fig. 1.3. Frontispiece to Gilbert White's *The Natural History and Antiquities of Selborne* (London: T. Bensley, 1789), plate 1. British Library, London.

list, preserved in his commonplace book held by The Huntington, includes the title. So did John Muir. The Huntington owns his inscribed copy.

The line from White to Muir, who wrote the last of his works on nature in the American West in the second decade of the twentieth century, is one followed in the trajectory of the *Storm Cloud* exhibition. Beginning with the Romantic poets and painters, the project traces the aesthetic responses that ran alongside the industrial revolution, tracking in different valences the profound shifts in everyday life that accompanied them: to people's relationship with time, with water, with locality, with weather, with plants and animals—wild and domestic. These relationships are the subjects of the novels, poems, essays, photographs, prints, drawings, and paintings explored in the exhibition and this publication. To begin to understand how they shifted, it is helpful again to turn to Gilbert White.

The frontispiece of White's book depicts a well-dressed group gathered on a hillside observing the landscape around them (fig. 1.3). They are shown in

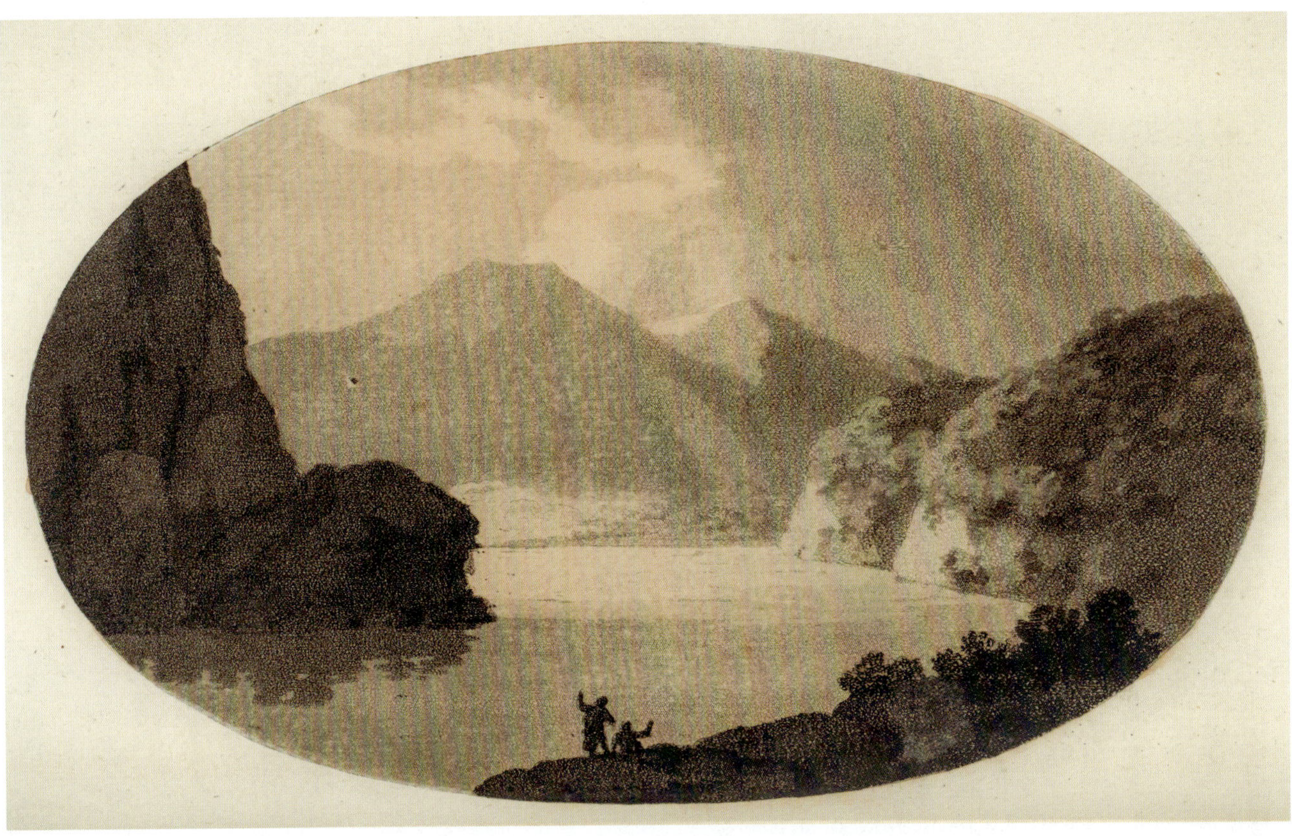

conversation; one gentleman raises his arm to point out the view to his companions. Their engagement with the landscape takes place beyond practical considerations. Nature, to them, holds an aesthetic value, a source of interest and pleasure to an individual observer. For the well-heeled group depicted in this engraving, nature is no longer what they live "in," but rather something they see, or must learn to see.

Fig. 1.4. William Gilpin, *Observations, Relative Chiefly to Picturesque Beauty, Made in the Year 1772, on Several Parts of England* (London: R. Blamire, 1786), facing p. 55. The Huntington, 606844.

By the late eighteenth century, a person's relationship with nature was beginning to be codified in new ways. Tourists, for example, sought out picturesque vistas for their enjoyment. This term, "picturesque," was applied to landscapes by the Reverend William Gilpin to refer to the kind of beauty found in a picture; in other words, nature that looked like art. This mode of viewing privileged age, variety, and notable natural features, and turned landscapes that included such things into sites of inspiration, as well as delight (fig. 1.4). Sublime landscapes—immense mountains, active volcanoes, deep chasms—offered the thrill of danger as another form of pleasure. The sublime was also an aesthetic category, first described by the Irish political philosopher Edmund Burke in 1757, and painters, like Joseph Wright of Derby, hinged their careers on visualizing the sensations it evoked. The sublime even posited industry as worthy of aestheticization.

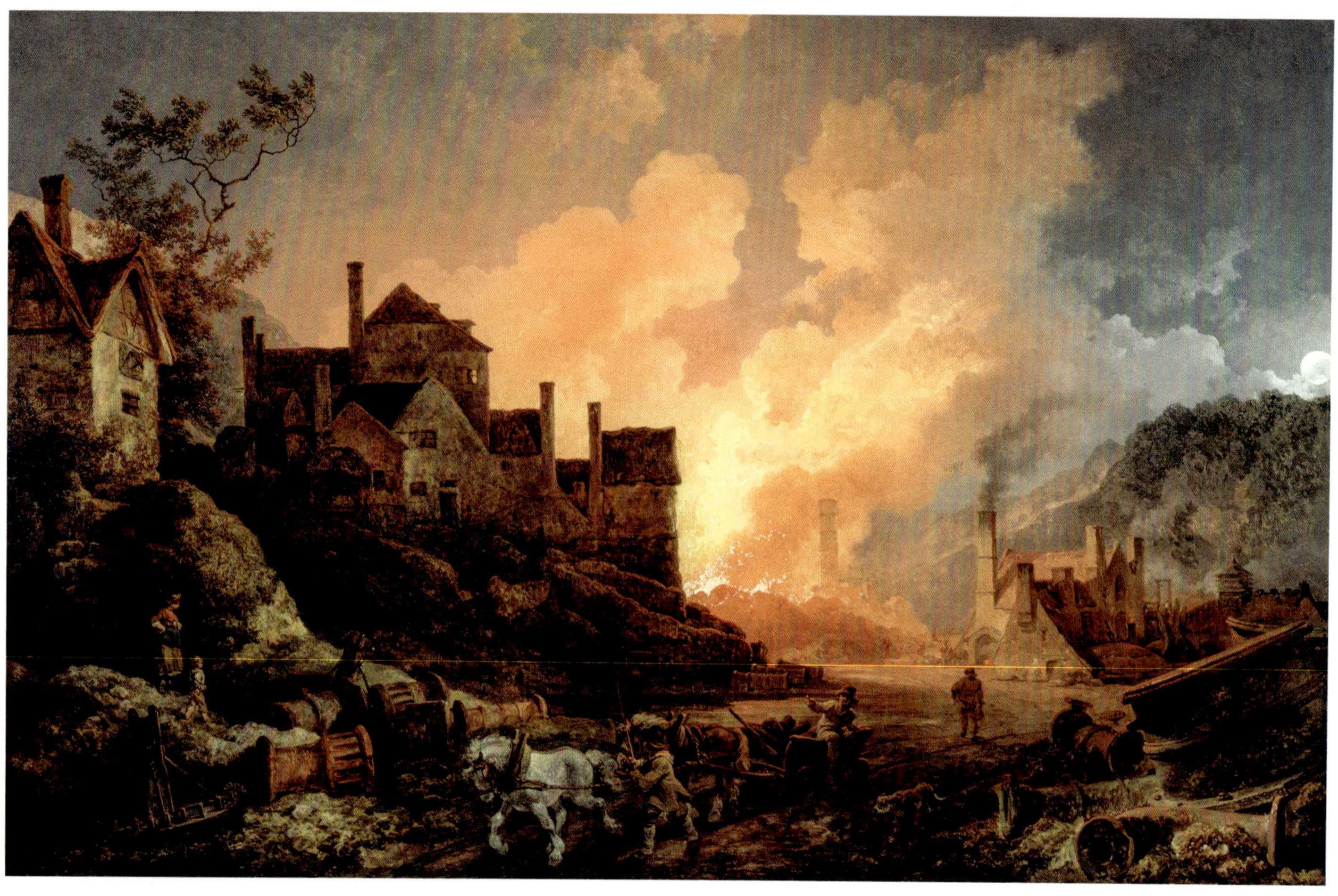

Fig. 1.5. Philippe Jacques de Loutherbourg, *Coalbrookdale by Night*, 1801. Oil on canvas, 26¾ × 42 in. (68 × 106.7 cm). Science Museum Group, 1952-452.

Philippe Jacques de Loutherbourg, for example, suggests a terrifying beauty in the heat of an ironworks' blast furnace at the same time he offers it as an experience for the tourist of picturesque landscapes (figs. 1.5, 1.6).

For millennia, literature has noted the seasons and mentioned local flora and fauna, but at the turn of the nineteenth century, a new approach to the description of nature in poetry arose, one perhaps best exemplified by the Romantic poet William Wordsworth. On daily countryside walks with his sister, Dorothy, and friends including Samuel Taylor Coleridge, Wordsworth sought to reintegrate with diurnal and seasonal changes and to live—and write—more simply. "Nature" was a guide, and daffodils and clouds as worthy of poetic description as political events and classical themes. Decades later, and in America, Transcendentalist writer Henry David Thoreau made a similar declaration when he moved into the cabin at Walden Pond, proclaiming, "I went to the woods because I wished to live deliberately."[4] His observations there form the basis for *Walden; or, Life in the Woods*, and his nature "Kalendar," which tracked blooming and leafing

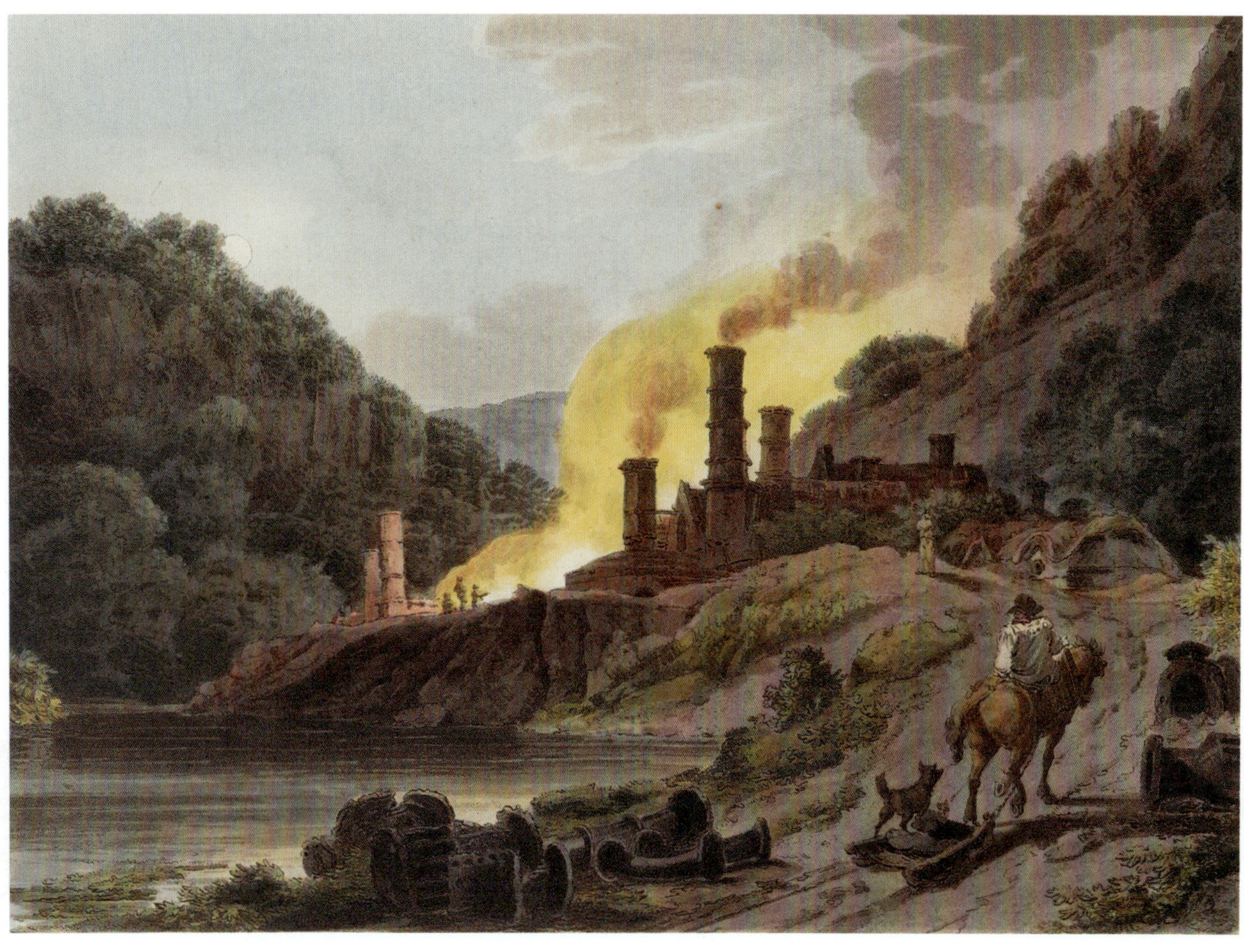

Fig. 1.6. "Iron Works of Coalbrook Dale" by William Pickett (engraver) after Philippe Jacques de Loutherbourg in *The Romantic and Picturesque Scenery of England and Wales* (London: R. Bowyer and T. Bensley, 1805), 36. The Huntington, 388992.

times. Thoreau kept this Kalendar over his final decade, seeking to construct a natural history of Concord, modeled on White's work on Selborne.

At the same time Wordsworth wandered "lonely as a cloud," Romantic painters were using their brushes to similarly describe the *experience* of nature as much as, if not more than, its appearance. When John Constable painted a ray of sunlight breaking through a cloud he expected us to know how it feels to witness the phenomenon (the momentary gleam of light hitting your eye; the feeling of the breeze that causes the cloud to move brushing against your skin), and what it means (a hope that the sun will soon come out). The painter famously chose to memorialize the more humble landscapes. Neither wild, nor urban, Constable's landscapes, marked by cornfields, mills, or canals, reflect the agricultural industries he grew up surrounded by (fig. 1.7). Like Wordsworth's, his is a personal and, above all, emotional relationship with the natural world, without which the conservation movements of later in the century would not have been possible. You have to love nature to want to preserve it.

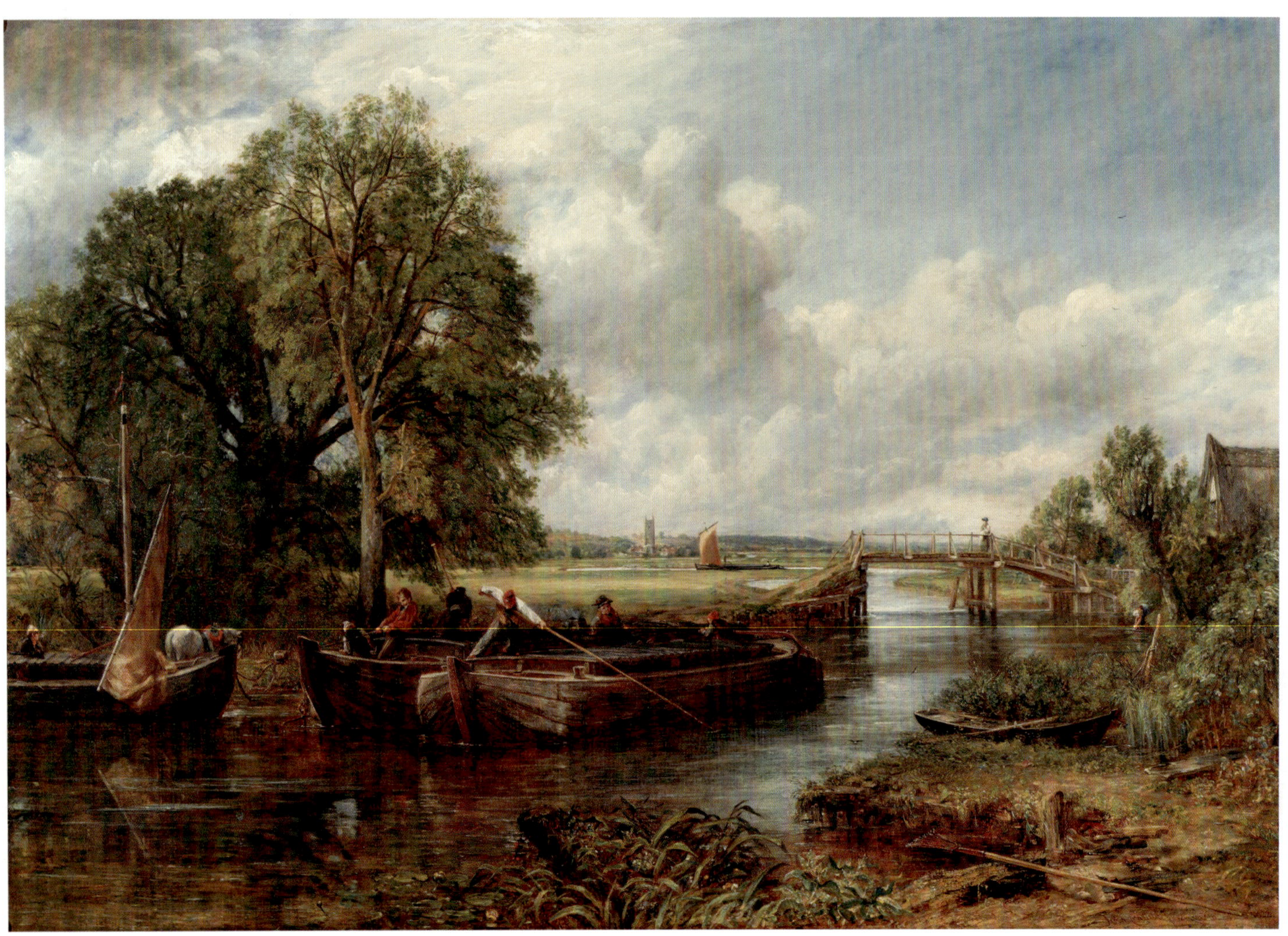

Fig. 1.7. John Constable, *View on the Stour near Dedham*, 1822. Oil on canvas, 51 × 74 in. (129.5 × 188 cm). The Huntington, 25.18.

Fig. 1.8. Mary Parker, Countess of Macclesfield, gentian and European mole crickets (?), in botanical sketchbook [9⅜ × 8¾ × 1⁵⁄₁₆ in. (23.8 × 22.2 × 3.3 cm)], watercolor on parchment, painting 19, 1756–67. The Huntington, mssHM 84100.

Just as landscape was being aestheticized for recreational experiences, the representation of natural specimens, such as plants and flowers, was entering the domestic realm by way of drawing masters, often versed in the latest science, who taught the art of botanical drawing to upper-class women and children as part of a genteel education. Their drawings, such as those by Mary Parker, include scientific accuracy as part of the aesthetic (fig. 1.8). Among those classes with the resources to engage in such activities, an appreciation of nature and the ability to respond to, and capture, its beauty were now part of an individual's cultural development.

In addition to drawing and painting natural specimens, many people assembled their own collections: pressed flowers, rocks and fossils, correlates to the displays in cabinets of curiosity that evolved into the natural history museums of the nineteenth century. A number of these "amateur" scientists amassed collections

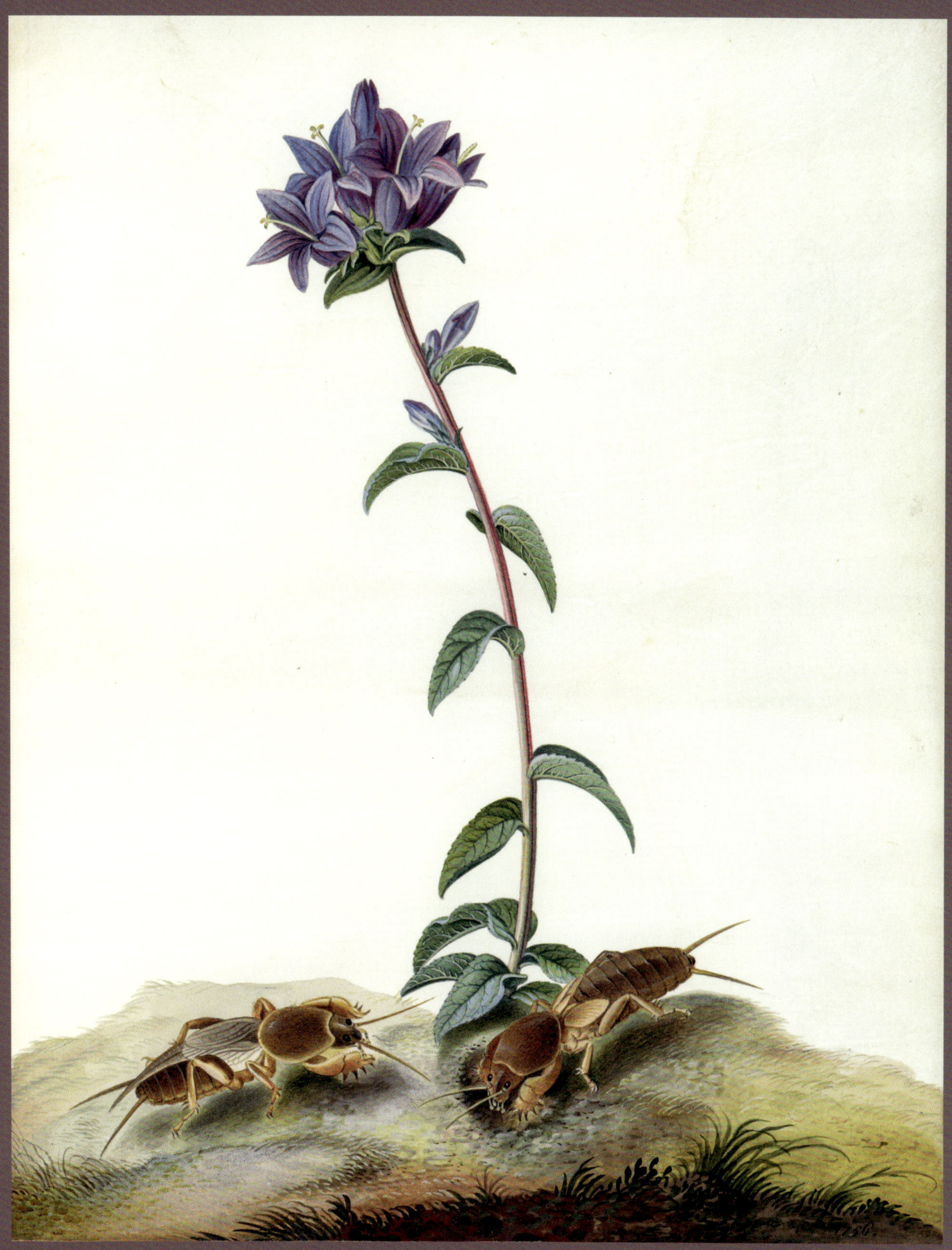

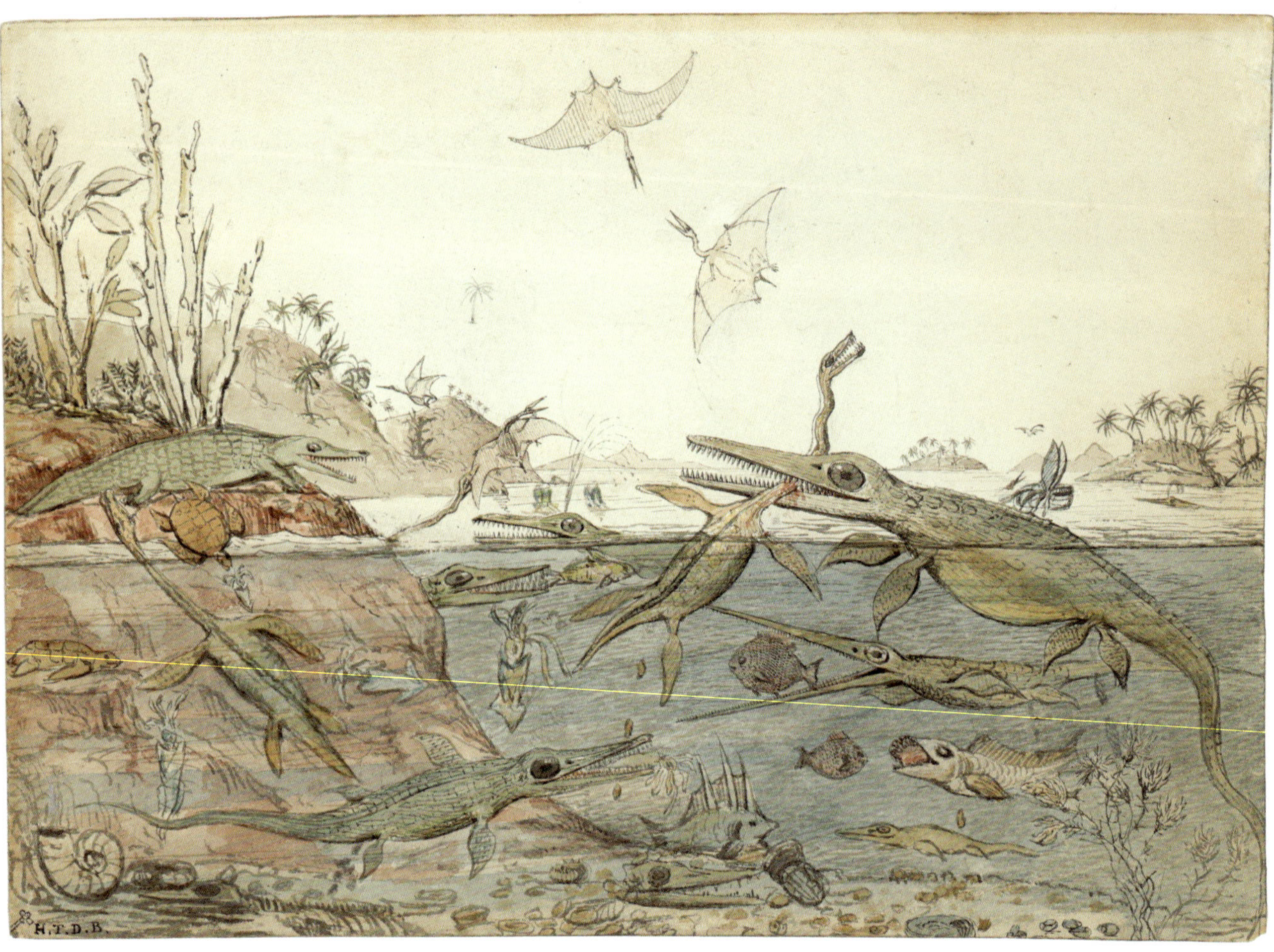

Fig. 1.9. Henry Thomas De la Beche, *Duria Antiquior*, 1830. Watercolor on paper, $9 \times 12\frac{5}{8}$ in. (23×32 cm). Courtesy of Amgueddfa Cymru—Museum Wales, Department of Geology, The De la Beche archive, 84.20G D368.

of great significance, and many of them were women. Mary Anning's fossil discoveries became the foundation of many paleontological collections around Britain, most significantly at what is now London's Natural History Museum. Though she found many of the fossils crucial to the early development of the field, Anning has only recently been credited as a pioneering paleontologist. In her lifetime, she struggled to make ends meet. Friend and fellow paleontologist, Henry Thomas De la Beche, painted a comedic scene illustrating her finds, which was reproduced and sold for her benefit (fig. 1.9). It is the first image of prehistoric life based on the fossil record.

While the arts were forging a new, more individual relationship between people and the natural world, an explosion of technological advancements brought about by the industrial revolution was altering the landscape itself through engineering projects and pollution, and taking people—who now worked in factories instead of on farms, and who now lived by clocks and train schedules instead of by the sun—further away from nature. If in 1780 the almanac was a primary tool for thinking with, and about, time in Europe and North America, by 1930 it had been replaced by the watch. In 1800, most people in Britain and the United States lived in the countryside versus the city. By 1900, the opposite was true in England, and nearly so in the United States.[5]

Now imagine the group from White's frontispiece visiting Hampstead Heath or closer to London's city center. Perhaps, for example, they are attending the "frost fair" held on the Thames in 1814, as seen in this image (fig. 1.10). How does nature appear in and register from within the city? And how is it differently valued from that vantage point? For centuries, the Thames periodically froze solid due to both seasonal reasons (cold winters during the Little Ice Age, 1300–1850)

Fig. 1.10. *A View of Frost Fair, on the Thames, February 1814*, 1814. Woodcut, overall: 17½ × 21¾ in. (44.5 × 55.2 cm). The Huntington, Print Box 980/8.

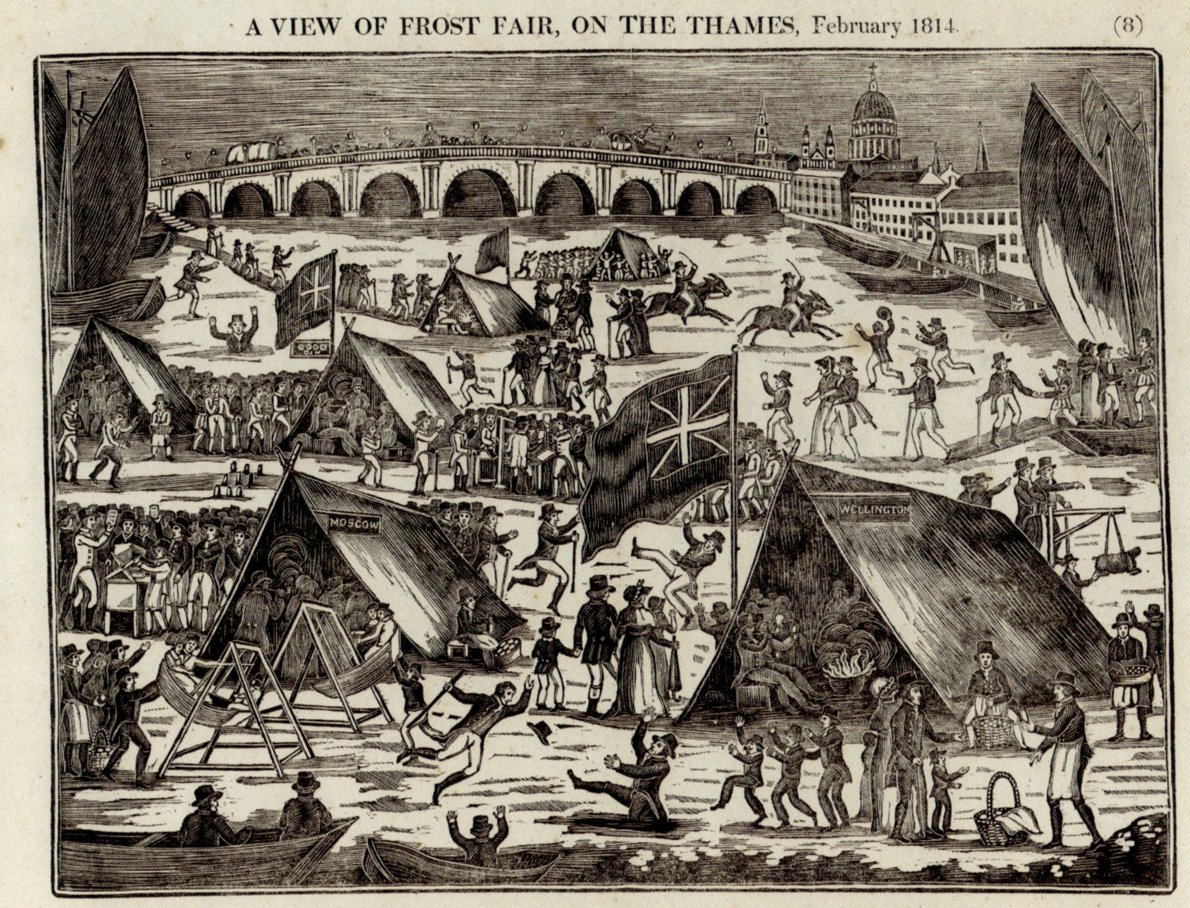

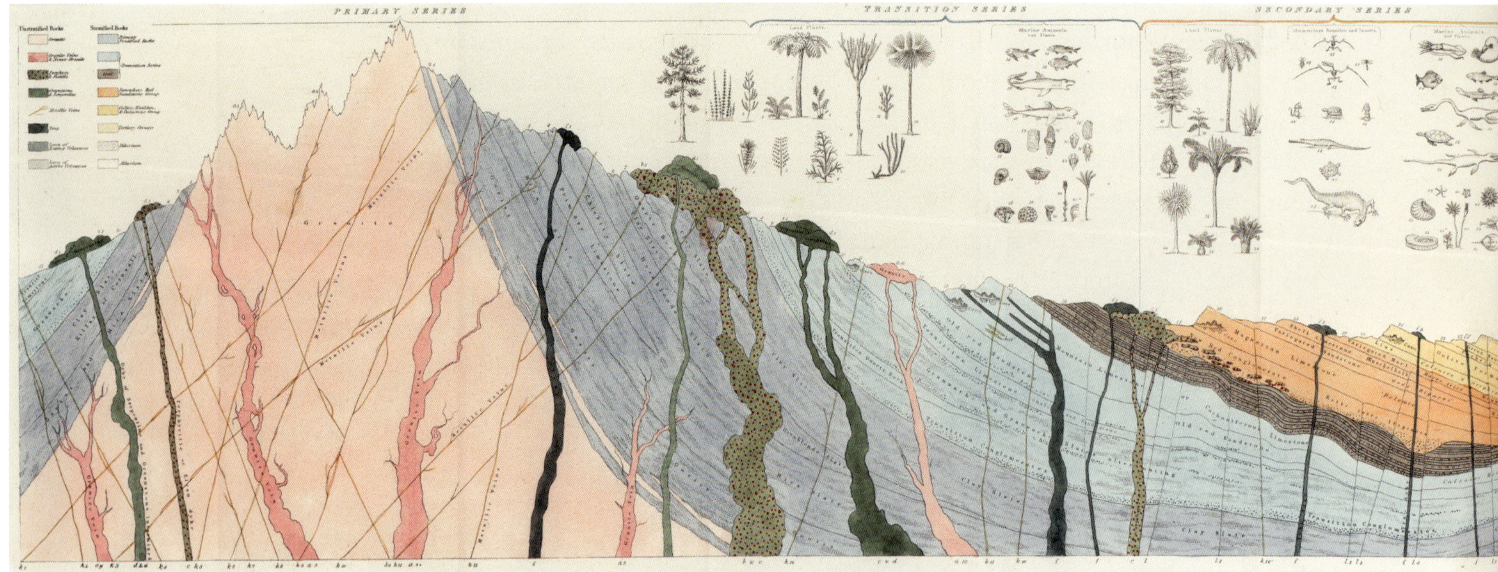

Fig. 1.11. "Ideal Section of a Portion of the Earth's Crust" in William Buckland's *Geology and Mineralogy Considered with Reference to Natural Theology* (London: W. Pickering, 1837), plate 1. The Huntington, 32526.

and infrastructural ones (London's bridges slowed the flow of the river). When the Thames did not freeze sufficiently to support a fair after 1814, scientists and London residents wondered why.

Emerging Sciences

For those keeping up with scientific developments, which were communicated in popular print magazines and lectures, the everyday experiences of time and nature weren't the only things that shifted during the nineteenth century. Only established as a field of study in the 1740s, geology upended humanity's conception of its place in Earth's history. In 1788, James Hutton wrote that "time, which measures every thing in our idea, and is often deficient to our schemes, is to nature endless and as nothing."[6] His *Theory of the Earth* and, later, Charles Lyell's *Principles of Geology* (1830–33) helped establish that the planet had existed for billions of years, a scale of "deep time" so much longer than biblical time, so much longer than recorded history, that the place of humans was radically reimagined. Another branch of geology, glaciology, showed that Earth's climate had changed over time, and many times, before the appearance of Homo sapiens. Glaciers remain important sites for the measurement and evocation of global warming. Long stratigraphic cross sections of mountains, reproduced on foldout engravings in books, introduced these new concepts to the public. William Buckland depicted various life-forms that existed over the course of Earth's history above a stratified landscape where their fossils might be found, vividly illustrating the concept that mass extinctions had occurred at multiple

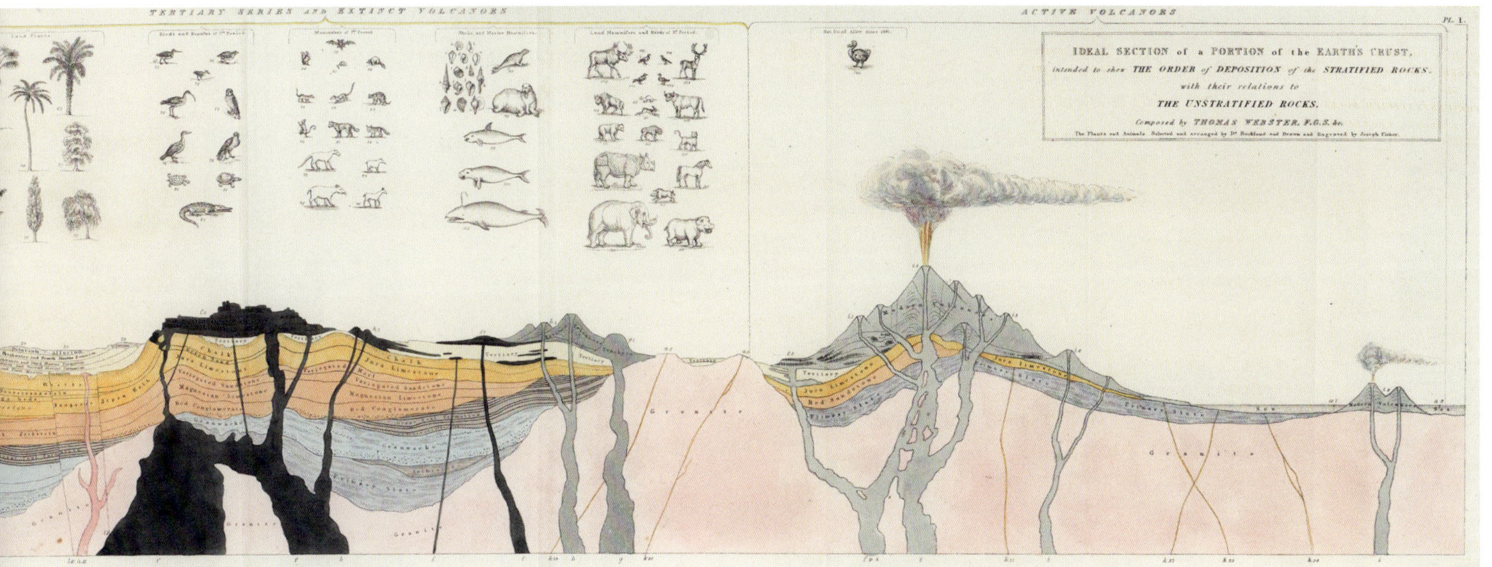

points in Earth's history, with a dodo at far right representing recently extinct animals (fig. 1.11 and detail, fig. 1.12).

Long concerned with predicting weather, meteorology in this period began to reflect methodologies adopted by other sciences and examine the effects natural events and human activities had on Earth's atmosphere. In his essay *On the Modifications of Clouds* (first published 1803), British chemist Luke Howard sought to depict and classify clouds by type (cirrus, cumulus, and so forth), much as Swedish botanist Carl Linnaeus had done for plants in the previous century. Later publications added further subdivisions to Howard's system of classification, and drew on a now international group of scientists, employing photography—which was especially well suited to capturing gradations of light and the quickly evolving "modification" or changeability of clouds—in projects such as the *Atlas international des nuages* (International cloud atlas) of 1896 (fig. 1.13). Howard's work had implications beyond meteorology; his *Climate of London* (first published 1818–20) established the first link between air pollution and temperature. Only a few decades later, in 1856, American scientist Eunice Foote became the first person to publish on carbon dioxide's greenhouse effect, forever implicating our use of fossil fuels (coal, then and since; petroleum for the last hundred or so years) in the degradation of our biosphere.[7]

Much as advances in the study of weather accompanied new understandings of the broader subject of climate, new approaches in natural history expanded its focus to consider the interactions among different systems: water, air, plants, animals, minerals. Naturalists such as Alexander von Humboldt embraced a holistic view of nature, asserting that the various scientific disciplines could

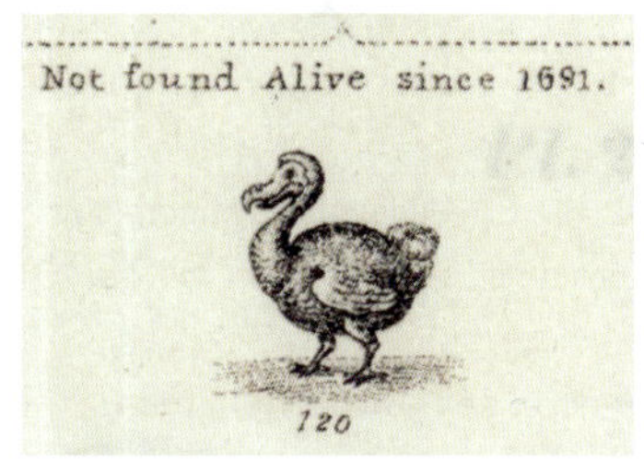

Fig. 1.12. Detail of dodo in "Ideal Section of a Portion of the Earth's Crust."

ALTO-CUMULUS

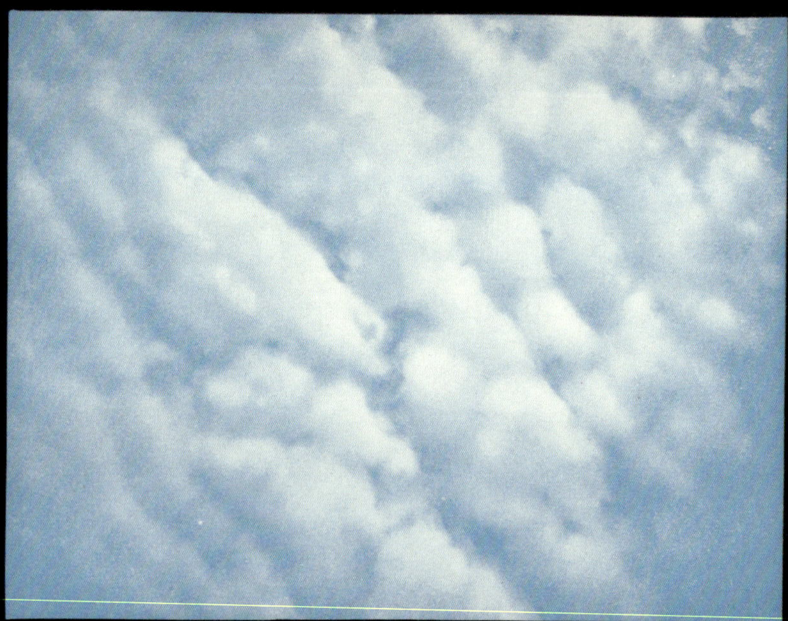

Fig. 9.

ALTO-CUMULUS

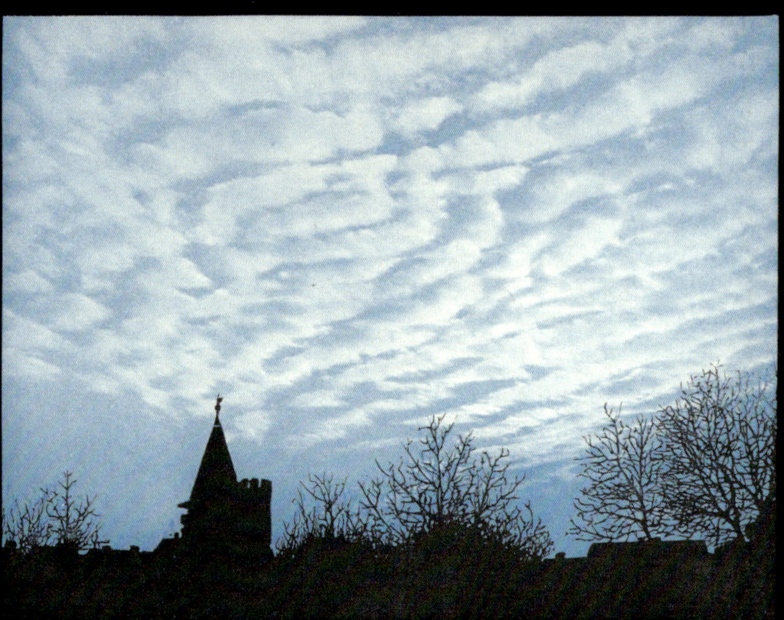

Fig. 10.

Photochromotypie Brunner & Hauser, Zürich.

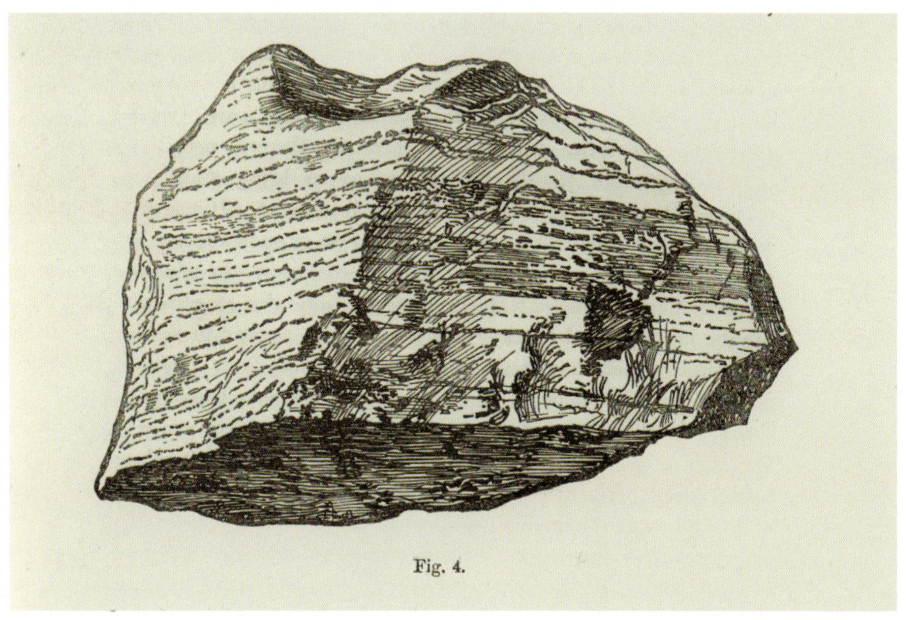

work together to reach a more complete understanding of the universe. Humboldt's description of the "unity of nature" utilized data from such disparate fields as biology, geography, and meteorology.[8] His work influenced countless others in the natural sciences, including Charles Darwin, and established many of the principles of the field we now know as ecology. At the same time, Humboldt's view of "exotic" nature, such as the landscapes of the Americas he so diligently described, engendered a belief that these lands existed outside of history, as an untouched, and therefore uninhabited, primordial world, open for exploration, and exploitation, though Indigenous peoples had occupied these places for thousands of years. This romantic conception of the American continent fostered not only expansionist doctrines like "manifest destiny" but also the countless sweeping vistas of majestic wilderness produced by nineteenth-century landscape painters (see, for example, pages 121 and 125).

Fig. 1.13. "Alto-Cumulus" in Hugo Hildebrand's *Atlas international des nuages* (International cloud atlas) (Paris: Gauthier-Villars et Fils, 1896), plate V.

Fig. 1.14. Piece of rock from the top of Mont Breven in John Ruskin's *Modern Painters*, vol. 4: "Of Mountain Beauty" (London: Smith, Elder, 1848–60), 109. The Huntington, 136782.

The Arts and Science

The practice of close and precise observation of nature unites the artists and scientists presented in this exhibition. Following approaches to naturalistic observation utilized by scientists such as Humboldt, John Ruskin collected, observed, and painstakingly drew and described not only clouds but also rocks and plants as he developed his theories of art and nature (fig. 1.14). His injunction that the artist depict nature directly and minutely "rejecting nothing, selecting nothing, and scorning nothing" had a profound influence on artists and writers,

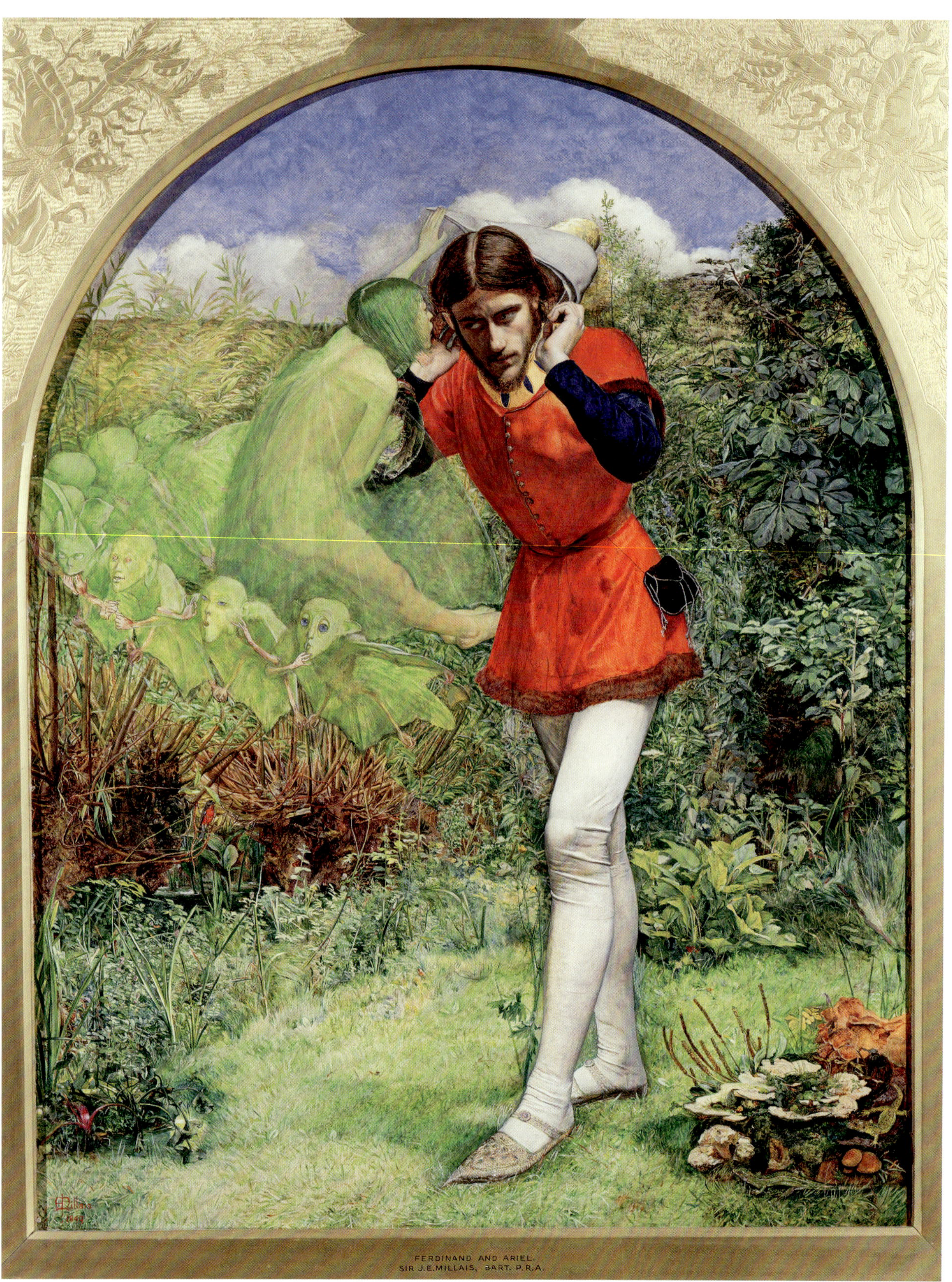
FERDINAND AND ARIEL.
SIR J.E.MILLAIS, BART. P.R.A.

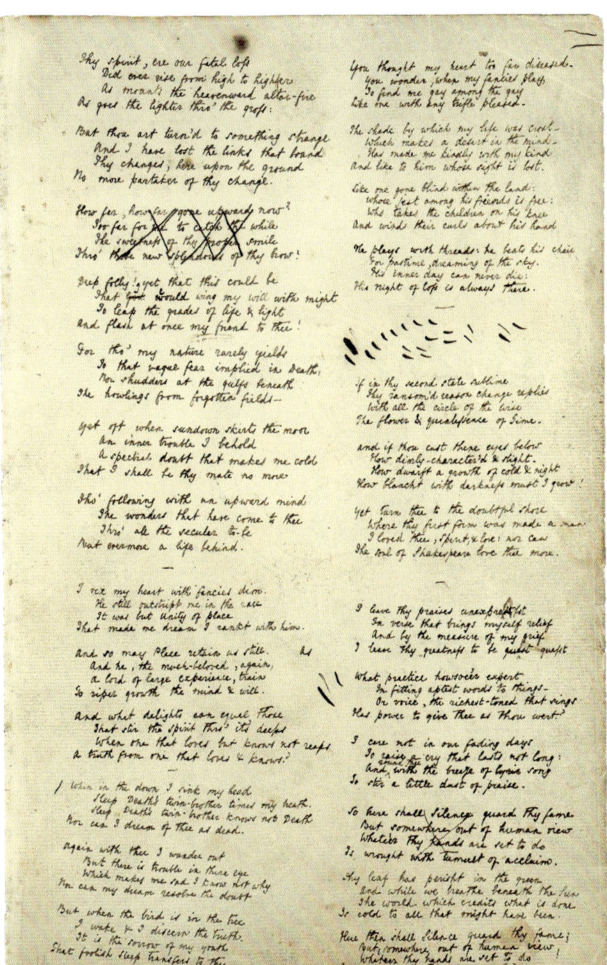

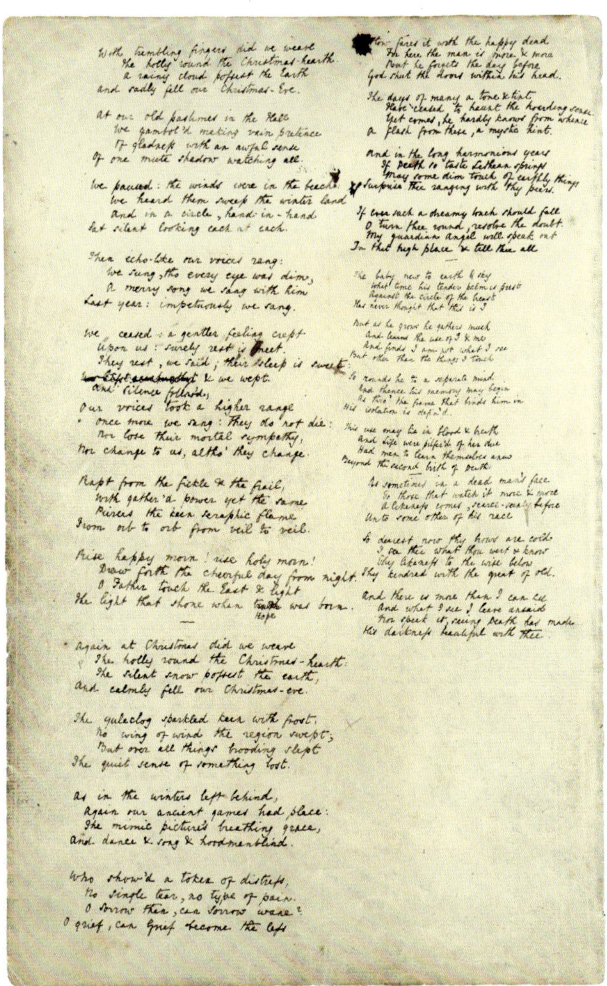

especially those of the Pre-Raphaelite circle, who depicted wild plants as they grew, not as a vague background to a biblical or literary scene, but with a specificity and precision that presaged the ecological thesis that all living species exist in a careful and complex balance.[9] We can see this web of nature in paintings like John Everett Millais's *Ferdinand Lured by Ariel* (fig. 1.15), where, among the Shakespearean characters, plants appear at various stages of life, some sprouting, some decaying, and act as habitat for other species.[10] For many of the artists and writers considered here, close observation of the natural world was an ethical as well as artistic project. Truth to nature encompassed a moral truth, one that is still relevant today.

In addition to our changing relationship with nature, the arts also tracked our changing relationship with time. In "In Memoriam," the poet Alfred, Lord Tennyson, lamented the altered relationship between humanity and the divine brought about by, among other things, new geological understandings of the age of the Earth (fig. 1.16): "I stretch the lame hands of faith, and grope, / And gather dust and chaff, and call / To what I feel is Lord of all, / And faintly trust the

Fig. 1.15. John Everett Millais, *Ferdinand Lured by Ariel*, 1849–50. Oil on panel, 25½ × 20 in. (64.8 × 50.8 cm). Private Collection/ Bridgeman Images.

Fig. 1.16. Alfred, Lord Tennyson, *In Memoriam* (partial draft), double-sided sheet of autograph manuscript, c. 1850. The Huntington, mssHM 1321.

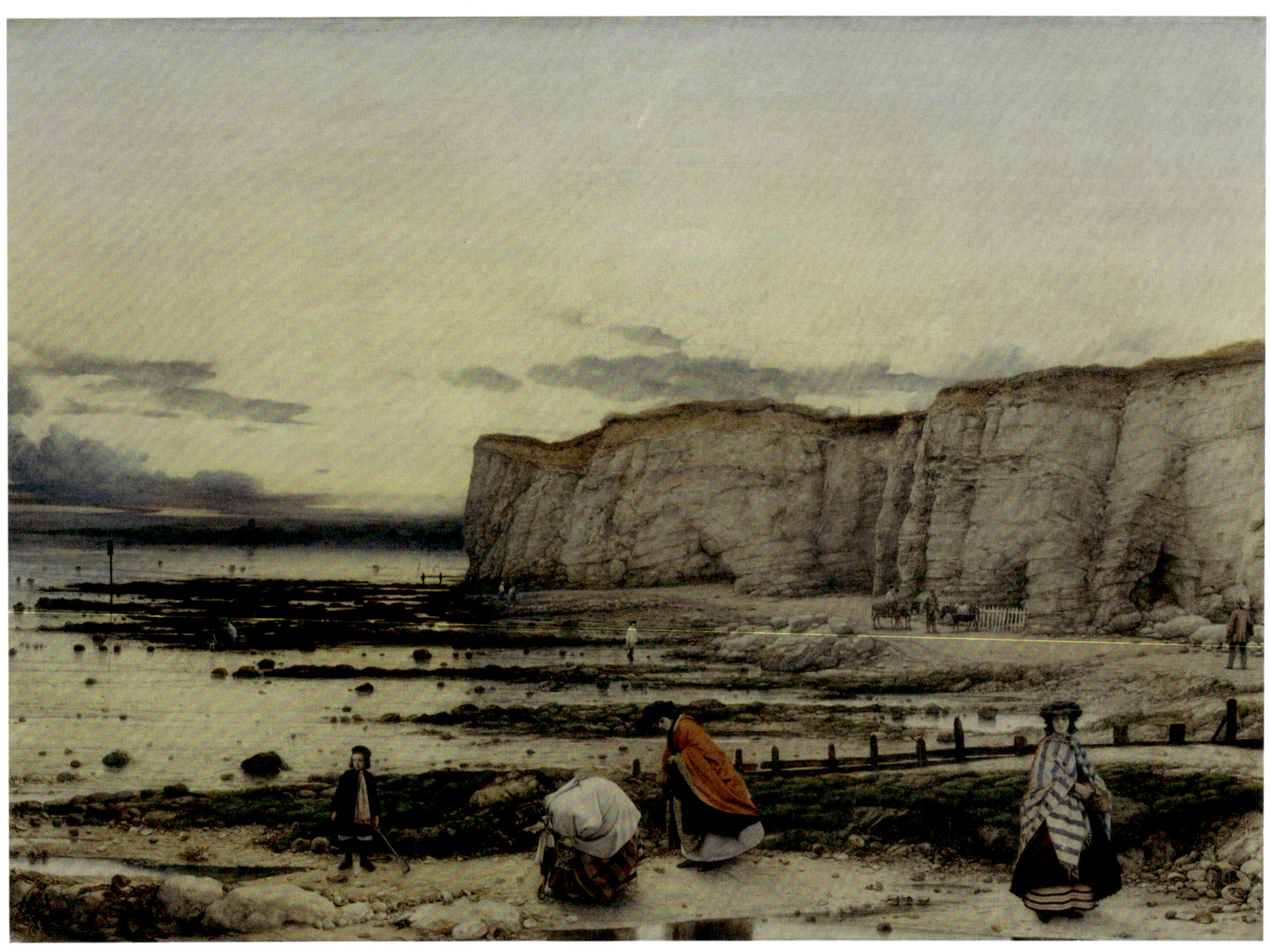

Fig. 1.17. William Dyce, *Pegwell Bay, Kent—A Recollection of October 5th 1858*, c. 1858–60. Oil on canvas, 25 × 35 in. (63.5 × 88.9 cm). Tate, London, purchased in 1894, N01407.

larger hope." William Dyce's painting *Pegwell Bay, Kent—A Recollection of October 5th 1858* explicitly references geology (in the detailed description of strata visible in the cliff), paleontology (in the family's hunt for fossils along the shore), and astronomy (in the depiction of Donati's comet in the sky) (fig. 1.17). Almost exactly contemporary with Darwin's publication of *On the Origin of Species*, the painting engages with developing scientific understandings of the profound histories of the Earth and stars. While some have read the disconnectedness of its figures, who do not interact with each other despite their common activity, as reflecting the artist's ambivalence toward humanity's newly revealed place in Earth's long history, others suggest the painting echoes contemporary thinkers who saw the beauty of rocks and vastness of space as evidence of God's existence.[11]

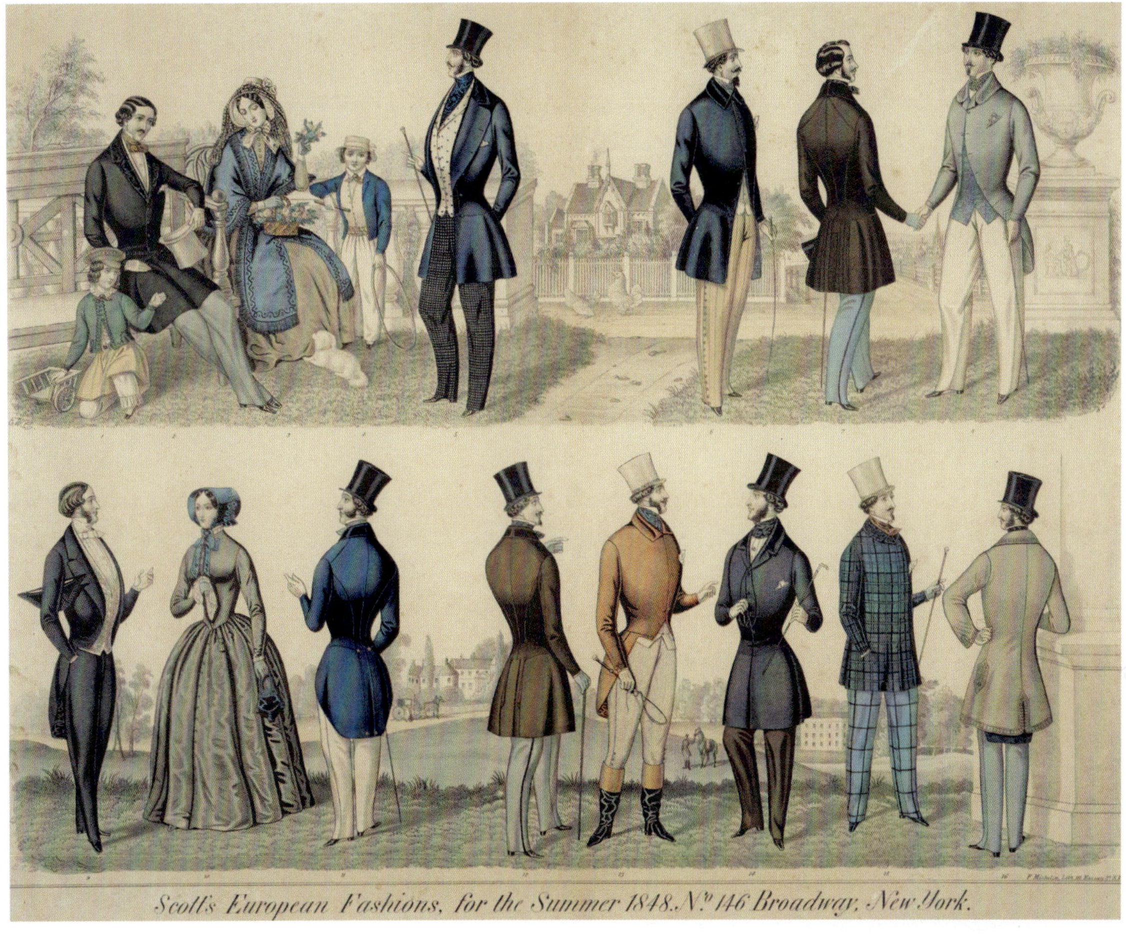

An understanding that mass extinctions had happened before, as with the dinosaurs, and could, or would, if the geological record was evidence, happen again, carried an ominous message for humans. The notion is explored in nineteenth-century speculative fictions, such as Mary Shelley's *The Last Man* (1826) and Richard Jefferies's *After London* (1885), which both described a post-apocalyptic world of greatly reduced population and civilization. By 1864, George Perkins Marsh expressed great anxiety about "the hostile influence of man" on the natural world.[12] In particular, he lambasted the practice of hunting birds and mammals for fashion. In his lifetime, multiple species of birds were hunted for their plumage to such extreme degrees that they went extinct early in the twentieth century. North American beaver populations were only able to recover once silk replaced pelts as the preferred material for top hats (fig. 1.18). Ironically, embodying both trends, prominent New Yorker John Jacob Astor, whose family's first fortune was made from beaver pelts, wrote a paleontologically informed science fiction novel, *A Journey in Other Worlds* (1894).[13]

Fig. 1.18. Francis Michelin, *Scott's European Fashions, for the Summer 1848. No. 146 Broadway, New York*, 1848. Lithograph with hand coloring, overall: 18¾ × 23⅜ in. (47.6 × 59.4 cm). The Huntington, Jay T. Last Collection, priJLC 001523.

Anthropogenic Problems

Storm Cloud's narrative grows out of The Huntington's collections, whose strengths in historical material from the British and American empires have suggested its focus on the English-speaking world and the Western Hemisphere north of the Equator—the parts of the world that industrialized most quickly during the period, and that have been the primary drivers of climate change. While the effects of globalized industrialization and trade reach everywhere, this show tracks the peak of the British Empire and the beginning of American ascendancy, as the two powers developed the industries, especially in manufacture and transportation, that not only fostered population growth but transformed how they would conduct war and expand their territorial reach. Many of these developments were undertaken with optimism about the changes they would effect, with the strident triumphalism of "manifest destiny" and other progress narratives.

The long nineteenth century, as charted here, is either firmly part of, or skirts the beginnings of, what is now often called the Anthropocene, a proposed geological era defined by human alteration of the natural systems and physical properties of the Earth. Scholars have proposed several dates for the beginning of the Anthropocene: the development of the planet's first cities and agriculture, the advent of the global market economy in the twelfth century, the beginning of Europe's industrial revolution at the end of the eighteenth century.[14] The Anthropocene Working Group recently offered a year between 1950 and 1954 as its start date (still to be determined at the time of this writing) because multiple measurements show that during these years human activities began to overwhelm natural systems.[15] However, the nineteenth century was a close contender for a start date due to environmental devastations wrought in the period, including pollution, the loss of biodiversity, impacts to ecosystems caused by deforestation and the conversion of wild lands to farmland, and steep increases in carbon dioxide.

Among humanists, the term "Anthropocene" is debated as well. It describes both the measurable effects of humans on the planet and the social, political, and economic factors that produce those changes. Some scholars argue that the "anthropos" (or "human") in "Anthropocene" suggests an evenly distributed culpability across all people. By this view, "Capitalocene" or "Plantationocene" would be more accurate because they hold certain economic and cultural agents most responsible for our current climate crisis.[16] *Storm Cloud* takes as a foundational concept that the story of the environment cannot be separated from the human condition, thus necessitating an exploration of practices, such as empire building, slavery, and factory labor, that were integral to the economies of the nineteenth century. Natural resources and human labor tend to be extracted and exploited concurrently.

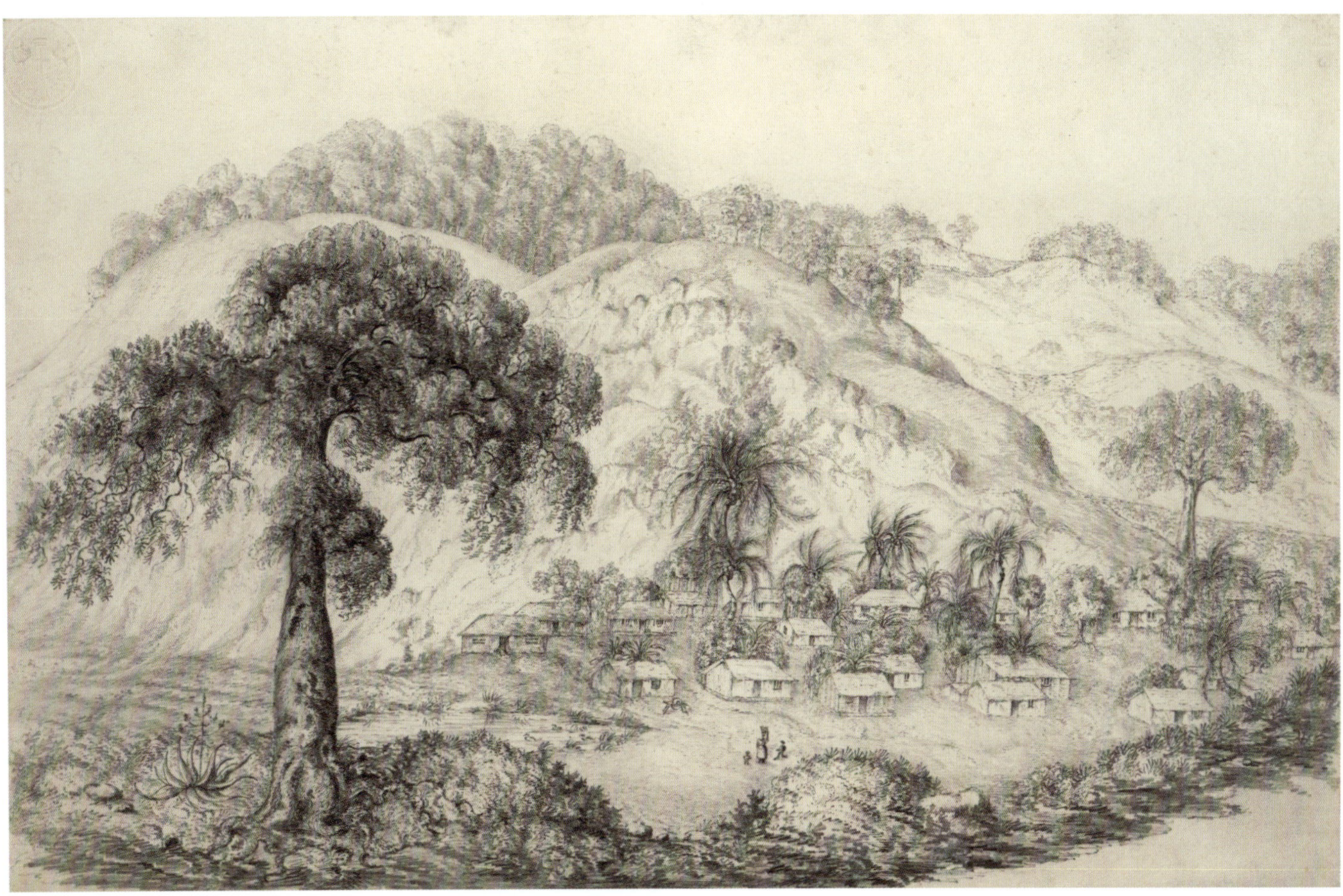

Fig. 1.19. Mary Clementina Barrett, *Slave Houses on the Barrett Plantation, Jamaica*, c. 1830. Graphite on embossed Turnbull's superfine board, 9 × 14 in. (22.9 × 35.6 cm). The Huntington, Purchased with funds from the Art Collectors' Council, with additional support from the Caillouette Acquisition Endowment for British and Continental Art, the Sara Smith Memorial Endowment, and the Robert R. Wark Art Acquisition Endowment, 2022.9.3.

We now understand that monocrop agriculture, the serial planting of a single species, depletes the soil's nutrients and leads to erosion. When it was introduced, export-oriented commodity agriculture supplanted local forms of "subsistence" agriculture that employed crop rotation and other methods to sustain the soil's fertility, and it did this on an increasingly industrial scale. In the nineteenth century, the work of enslaved people made monocrop agriculture not only possible but also very profitable. Written and visual descriptions of Jamaica, for example, reveal how the plantation system affected both people and the environment. The system had a dehumanizing effect, as seen in a drawing in the Huntington collection by Mary Clementina Barrett, wife of a British enslaver, in which she records the local plant life and enslaved laborers of her Jamaican home as if they were all part of the same inherent structure, witness to her acceptance and even naturalization of the colonial system that generated her family's wealth (fig. 1.19).

While many technological advancements brought convenience or prosperity to certain sectors, some questioned whether they introduced more misery than

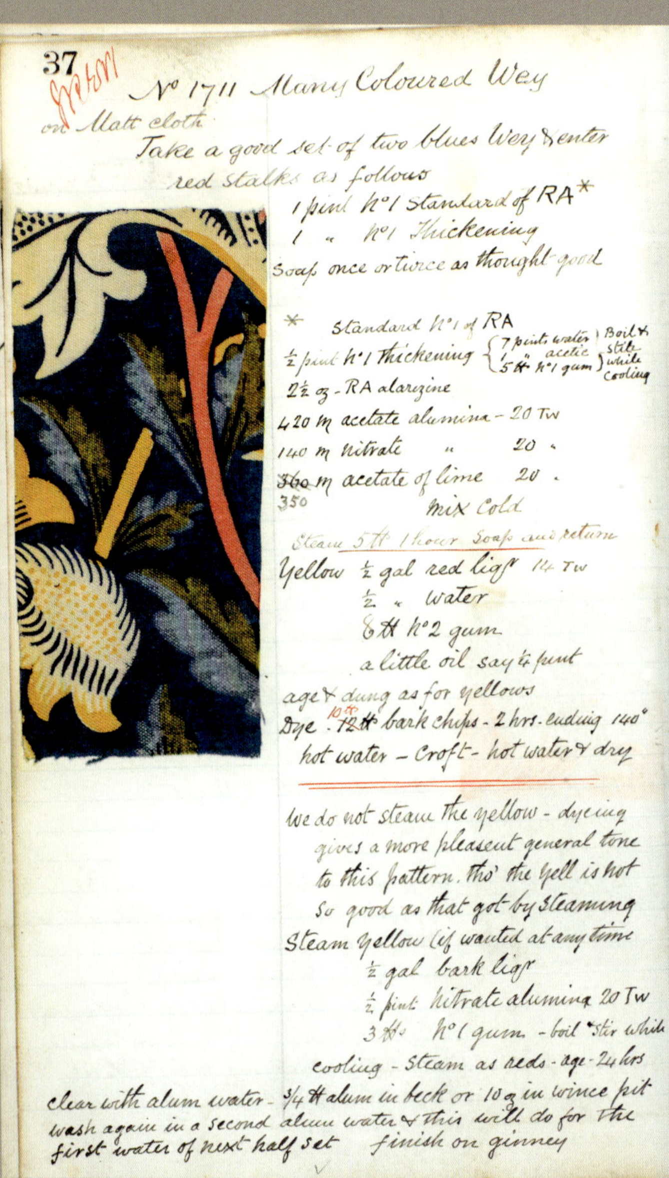

37

No 1711 Many Coloured Wey
on Matt cloth
Take a good set of two blues Wey & enter
red stalks as follows
1 pint No 1 Standard of RA*
1 " No 1 Thickening
Soap once or twice as thought good

* Standard No 1 of RA
½ pint No 1 thickening {7 pints water, 1 " acetic, 5 lb No 1 gum} Boil & stir while cooling
2½ oz - RA alarzine
420 m acetate alumina - 20 Tw
140 m nitrate " 20 "
360 m acetate of lime 20 "
350
mix cold
Steam 5 lb 1 hour Soap and return
Yellow ½ gal red liqr 14 Tw
½ " water
8 lb No 2 gum
a little oil say ½ pint
age & dung as for yellows
Dye: 12 lb bark chips - 2 hrs. ending 140°
hot water - Croft - hot water & dry

We do not steam the yellow - dyeing
gives a more pleasant general tone
to this pattern, tho' the yell is not
so good as that got by steaming
Steam yellow (if wanted at any time
½ gal bark liqr
½ pint nitrate alumina 20 Tw
3 lbs No 1 gum - boil & stir while
cooling - Steam as reds - age - 24 hrs
clear with alum water - 3/4 lb alum in beck or 10 oz in wince pit
wash again in a second alum water & this will do for the
first water of next half set   finish on ginney

38

No 1494 Loddon on Matt cloth
This is on a blue dyed purposely & known as Loddon blue
Print - catechu as blue windrush fol 32
Discharge - 1 quart thickening
2 oz oxalic in ¼ pint water
Thickening = 1 gal water 2 lb No 2 gum
4 lb No 3 gum 4 lb clay
clear as blue Windrush and return for red
Red 1 qt No 1 thickening {7 pts water, 1 " acetic, 5 lb No 1 gum} boil & stir while cooling
1½ oz medium aliz
4 3½ oz acetate of alumina 20° Tw
1 oz 160 m nitrate 20
2 oz 440 m acetate of lime 20
3 oz 160 m
8 drs Tin crystals
Pink 2 qts No 1 thickening (as above for red)
See below for Pink
20 oz medium alizarine
7 oz acetate of alumina 20 Tw
2 oz 320 m nitrate 20
5 oz 400 m acetate of lime 20
To 1 pint of this standard add
4 pints No 1 thickening
10 oz acetic 10 oz water
Steam & clear as other alizarines 5 lb 1 hour
2nd Soap at 160° and return for yell
Yellow - mordant 1/2 gal red liqr 14 Tw
1/2 " water
8 lb No 2 gum 20 oz Starch
20 oz dragon
dye 40 lb welds at once or
30 lb & 15 lb at twice?
Pink - ½ pint ... Red Colour & add
1½ pints No 1 Thickening
½ oz acetic

Fig. 1.20. John Henry Dearle, *Merton Abbey Dye Book*, c. 1882–91, pages 37–38. The Huntington, 2000.5.3387.

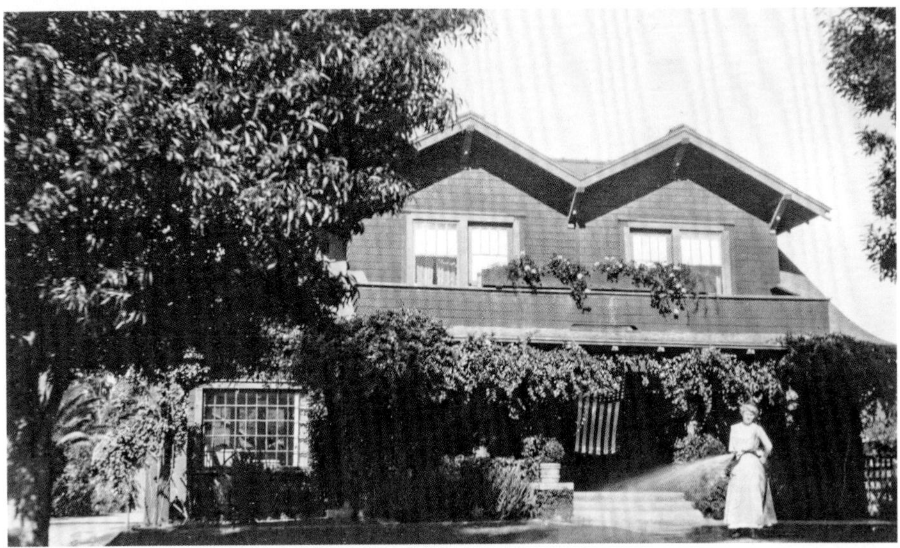

Fig. 1.21. "1837 Canyon Drive, and the Slacker Using the Water" in William H. Frick and Julius G. Oliver's "The Los Angeles Aqueduct," 1917, folio 30. The Huntington, photCL 442 (30).

they prevented. The original "muckraker," Upton Sinclair, perhaps best known for his exposé of the meat-packing industry, also wrote about coal and oil. The introduction of mechanized looms in England's textile factories provoked the Luddite worker rebellions of the early nineteenth century, which protested the accompanying loss of jobs and reduction of wages. Factory labor produced not only the air pollution railed against by Sinclair and Ruskin, but also the degrading working conditions that inspired people like William Morris to attempt a return to preindustrial practices. In the decorating firm he established in 1861, Morris emphasized both handicraft and the dignity of its labor, imagining that art, including the furnishings with which one decorated one's home, was made in ages past "by the people for the people as a joy for the maker and the user."[17] His firm revived the use of vegetable dyes, which had been replaced by modern chemical dyes that polluted nineteenth-century waterways, their formulas prescribed in a recipe book used in the production of printed textiles at his Merton Abbey workshop, now preserved in The Huntington's collection (fig. 1.20).

From Then to Now

The environmental consequences noted by observers in the nineteenth century, only some of which were recognized as problems, are our legacy. The Huntington owns a photographic album, made near the end of the period covered by this exhibition, celebrating one of the engineering projects that brought water to the Los Angeles region, the aqueduct of 1913. The final photograph in the album depicts a woman standing in the yard of her Craftsman home holding a garden hose spouting water (fig. 1.21). This image of bountiful water available for home gardening might read as ironic to Californians who have experienced conservation restrictions placed on water use because of years of drought. The

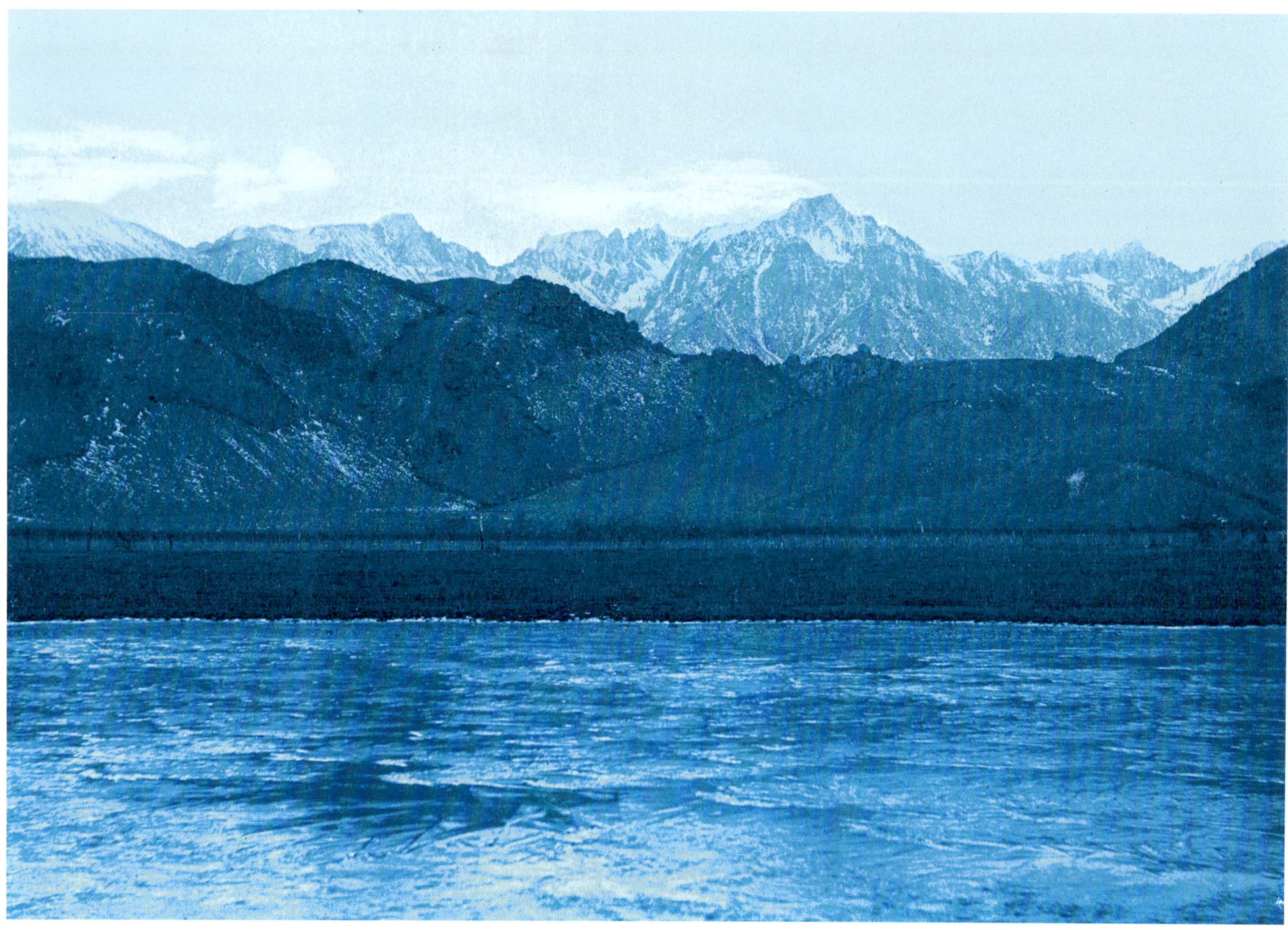

Fig. 1.22. *Owens Lake*, c. 1910. Cyanotype, 7 × 9 in. (17.8 × 22.9 cm). The Huntington, Mary Hunter Austin papers, 1845–1950 (bulk 1920–1934), photCL 296 fld. 5 (13).

1913 project had more serious implications to the Indigenous people of the Owens River Valley, whose water was diverted to Los Angeles. A pair of historic photographs showing Owens Lake before and after the aqueduct makes starkly visible the kinds of losses imposed upon California's Indigenous populations by policies that viewed their rights as secondary to the needs of industry and urbanization (figs. 1.22, 1.23).

What *Storm Cloud*'s historical arc only suggests through a dotted line, one initiated by the early twentieth-century shift from a primarily coal-based to a primarily petroleum-based economy, is the world that we live in today. This world has additionally registered the impact of automobiles, asphalt, plastics, and nuclear weapons, and had its view of the Earth complicated by space exploration. The last one hundred years have witnessed an acceleration of both the market and the mechanical (and now digital) technologies that have exponentially ramped up the climate crisis.

Contemporary critic Donna Haraway asks people to reckon with the Anthropocene and to "stay with the trouble," a mode of engagement that avoids the

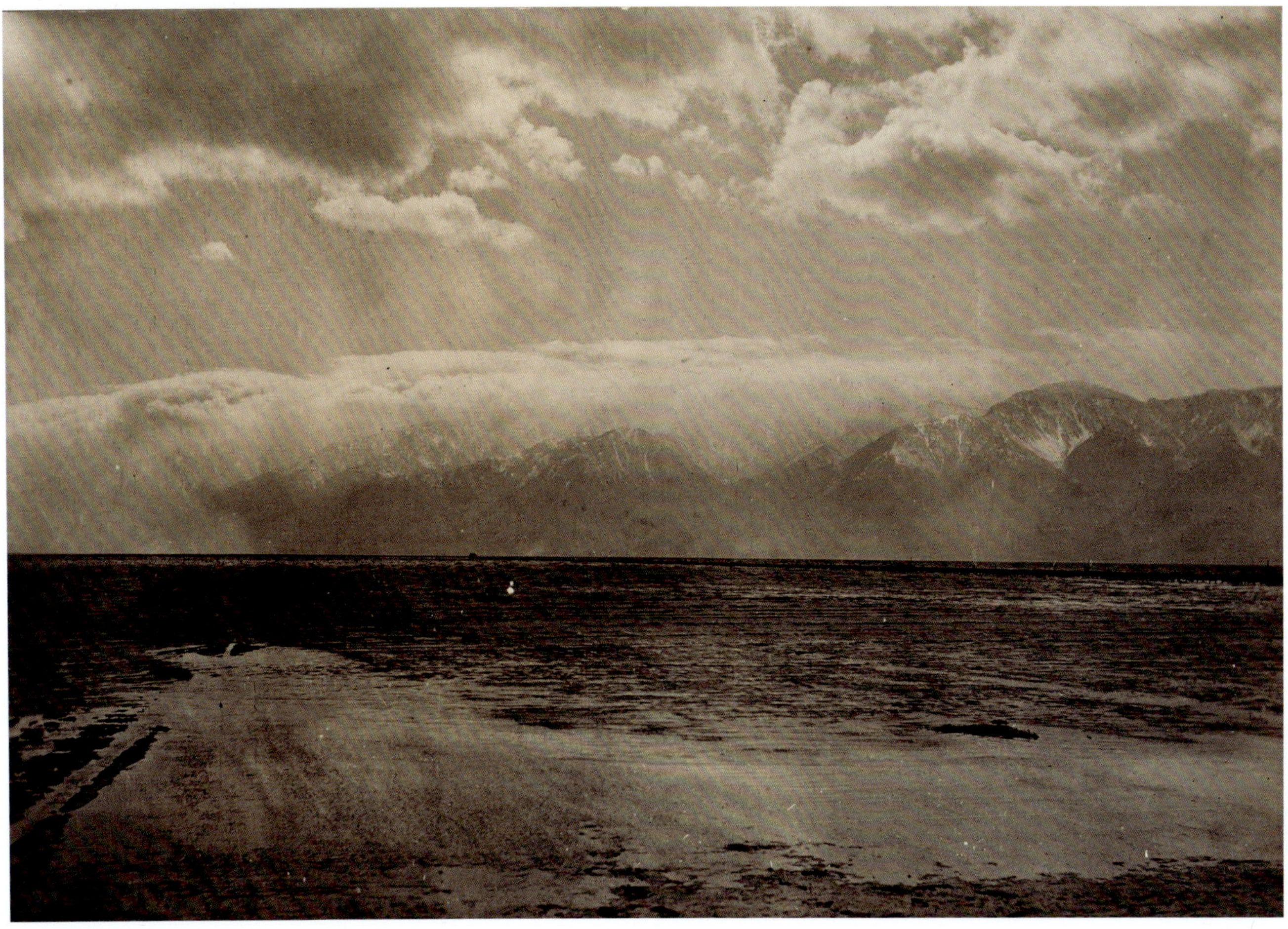

Fig. 1.23. *Owens Lake*, c. 1920. Gelatin silver print, 7 × 8⅞ in. (17.8 × 22.5 cm). The Huntington, Mary Hunter Austin papers, 1845–1950 (bulk 1920–1934), photCL 296 fld. 5 (11).

pitfalls of false optimism, apathy, or despair.[18] Addressing the climate crisis is challenging because the possible outcomes are both terrifying and unknown. Modern life has arisen through the industrial infrastructures and practices explored in this volume, a way of living from which it is now difficult to extricate—at practical, political, and imaginative levels. But the *Storm Cloud* exhibition demonstrates that this is not a new problem. Ruskin begins his "Storm-Cloud" lectures noting the dilemma, a phenomenon he saw as "peculiar to our own times."[19] Through Ruskin and others—scientists and artists, and many who were both or somewhere in between—we trace a set of observational practices across the sciences and literary and visual arts of the nineteenth century. These practices allowed them to register the impacts of looming environmental crisis on everyday life even before people were able to diagnose causes or suggest solutions. They ask us to consider what world we have made, what we have lost, what our "surround" is now, and how we might learn to see and experience it anew.

This book, and the exhibition it accompanies, is offered with the hope that establishing some historical context will help us stay with the trouble. It explores a mode of seeing, studying, and depicting nature as a means to express awareness of (and in some cases mounting concern over) environmental changes across a wide range of art and literature during the long nineteenth century. It is not a traditional exhibition catalogue. Two extended thematic essays and a series of short essays on individual objects speak to the deeply intertwined historical dialogue between art and science. The contributors represent a broad spectrum of academic inquiry and professional expertise—not only art and literary historians but also scientists, artists, activists, and representatives of Indigenous knowledge—individuals representing a range of stakeholders in our current moment of environmental reckoning. By engaging writers from a variety of perspectives to analyze the cultural legacy of the nineteenth century as it relates to the present day, this volume aims to lend a crucial multivocality to conversations around environmental justice and the historical basis of our current climate crisis.

Melinda McCurdy is Curator of British Art at The Huntington.
Karla Nielsen is Curator of Literary Collections at The Huntington.

Notes

1. See Nicholas Robbins, "Ruskin, Whistler, and the Climate of Art in 1884," in Kelly Freeman and Thomas Hughes, eds., *Ruskin's Ecologies: Figures of Relation from "Modern Painters" to "The Storm-Cloud,"* Courtauld Books Online, https://doi.org/10.33999/2021.56; and Jesse Oak Taylor, "Storm-Clouds on the Horizon: John Ruskin and the Emergence of Anthropogenic Climate Change," *19: Interdisciplinary Studies in the Long Nineteenth Century*, no. 26, https://doi.org/10.16995/ntn.802. Ruskin and climate change has also been explored in Suzanne Fagence Cooper and Richard Johns, *Ruskin, Turner, and the Storm Cloud* (London: Paul Holberton Publishing, 2019).
2. Gilbert White, Frederick Marns, and Thomas Bewick, *The Gilbert White Museum Edition of "The Natural History of Selborne"* (1900; repr., London: Shepheard-Walwyn, 1977), 104.
3. Steph Holt, "Gilbert White: The Modern Naturalist," Natural History Museum[, London]; https://www.nhm.ac.uk/discover/gilbert-white.html.
4. Henry David Thoreau, *Walden* (Boston: Beacon Press, 2017), 85.
5. Eric Hobsbawm, "Table 2: Urbanisation in Nineteenth-Century Europe, 1800–1890," in *The Age of Empire, 1875–1914* (New York: Vintage, 1989), 343. "Urbanization over the Past 500 Years, 1500 to 2016," Our World in Data, http://www.ourworldindata.org/grapher/urbanization-last-500-years.
6. James Hutton, *Theory of the Earth*, e-version on Project Gutenberg, 1795/2004, https://www.gutenberg.org/cache/epub/12861/pg12861-images.html.
7. Eunice Foote, "Circumstances Affecting the Heat of the Sun's Rays," *American Journal of Science and Arts*, 2nd ser., vol. 22 (1856): 382–83.
8. Alexander von Humboldt, *Views of Nature; or, Contemplations on the Sublime Phenomena of Creation*, trans. Elise C. Otté and Henry G. Bohn (London: H. G. Bohn, 1850), 285.
9. John Ruskin, *Modern Painters*, 3rd ed. (London: Smith, Eder, 1846), 1:417.
10. See John Holmes, *The Pre-Raphaelites and Science* (New Haven: Yale University Press, 2018), 19–43.
11. Christiana Payne, "*Pegwell Bay, Kent—A Recollection of October 5th 1858* ?1858–60 by William Dyce," esp. "Art, Science, and Religion," https://www.tate.org.uk/research/in-focus/pegwell-bay-kent-william-dyce/art-science-religion.
12. George Perkins Marsh, *The Earth as Modified by Human Action, a New Edition of "Man and Nature"* (New York: Scribner, Armstrong, 1874), 34.
13. Horace T. Martin, *Castorologia; or, The History and Traditions of the Canadian Beaver* (Montreal: W. Drysdale, 1892), 58–59.
14. Simon L. Lewis and Mark A. Maslin, "Defining the Anthropocene," *Nature* 519, no. 7542 (March 2015): 175, https://doi.org/10.1038/nature14258.
15. Associated Press, "Scientists Say New Epoch Marked by Human Impact—the Anthropocene—Began in 1950s," July 11, 2023, https://www.npr.org/2023/07/11/1187125012/anthropocene-crawford-lake-canada-beginning.
16. Donna Haraway, "Anthropocene, Capitalocene, Plantationocene, Chthulucene: Making Kin," *Environmental Humanities* 6, no. 1 (May 1, 2015): 159–65, https://doi.org/10.1215/22011919-3615934.
17. Quoted in Diane Waggoner, ed., *The Beauty of Life: William Morris and the Art of Design* (New York: Thames and Hudson, 2003), 9.
18. Donna J. Haraway, *Staying with the Trouble: Making Kin in the Chthulucene* (Durham, NC: Duke University Press, 2017), 2.
19. John Ruskin, "The Storm-Cloud of the Nineteenth Century," 1884, e-version on Project Gutenberg, https://www.gutenberg.org/files/20204/20204-h/20204-h.htm.

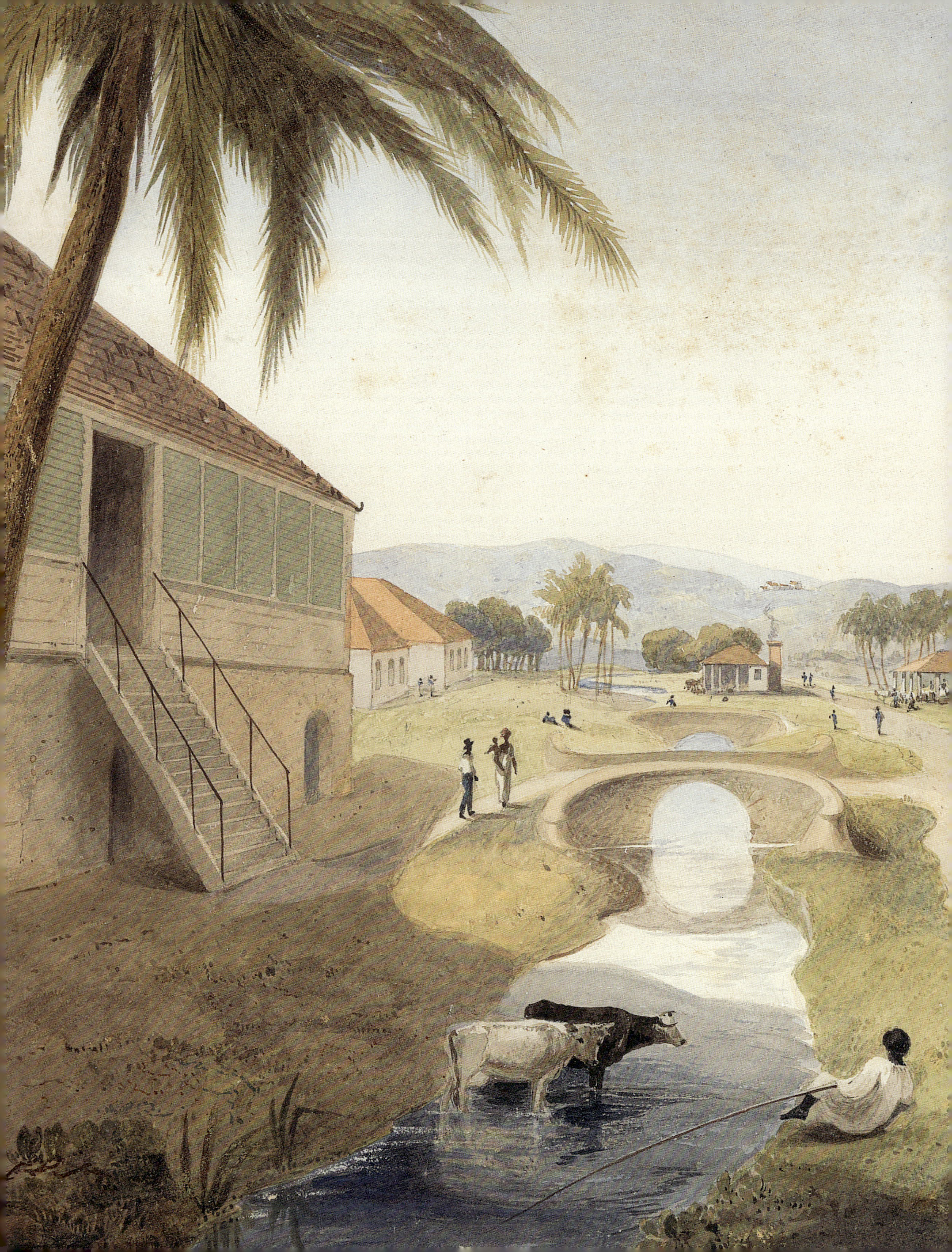

Nicholas Robbins

Forms of Water

Infrastructure and Improvement in the Nineteenth Century

Two versions of abundance from California's San Joaquin Valley in the late nineteenth century: one is a photograph made by Charles C. Pierce of a Yokuts weaver, her name not recorded, living on the Tule River Reservation (fig. 2.1). She is working on a large, coiled basket, her hands manipulating the grasses and roots that will be woven into the gentle curves of its fibrous body, which is encircled by two bands of linked human figures and rattlesnake designs. The other is a photograph made by Carleton Watkins of an artesian well expelling what its caption calls "cool, clear, crystal water," propelled into the air from aquifers deep below the earth (fig. 2.2). The well's spray forms a figural, almost living object whose translucent matter catches the energy of strong sunlight behind it. Pierce's photograph shows us the slow work of weaving—perhaps the making of a cooking basket, which would catch and retain water so that it can be heated with fire-baked stones and used to cook meals. Such baskets' negative, capacious forms aim to maximize water's capacity and its sustenance of life. Watkins's photograph is an image of water being instantaneously transfigured into an object. It is intended for growing food, but that use is only implicit. Water rushes toward the edge of the frame, suggesting the endlessness of resources: "never failing," "perpetual," the photograph's caption insists.[1]

This essay traces how water and its uses shaped the imagination of landscape and environment during the nineteenth century. It is concerned with how infrastructures and images were both used to transform water into an object that might be made to *work*. The essay will end in the same California valley where these two responses to natural resources originated. Taken together, they bring us immediately into contact with the political and social dimensions of environmental history. The spray of water seen in Watkins's photograph signals settler-colonial forms of land expropriation, technological optimization, and resource-intensive commodity production. Watkins's record of capitalism's

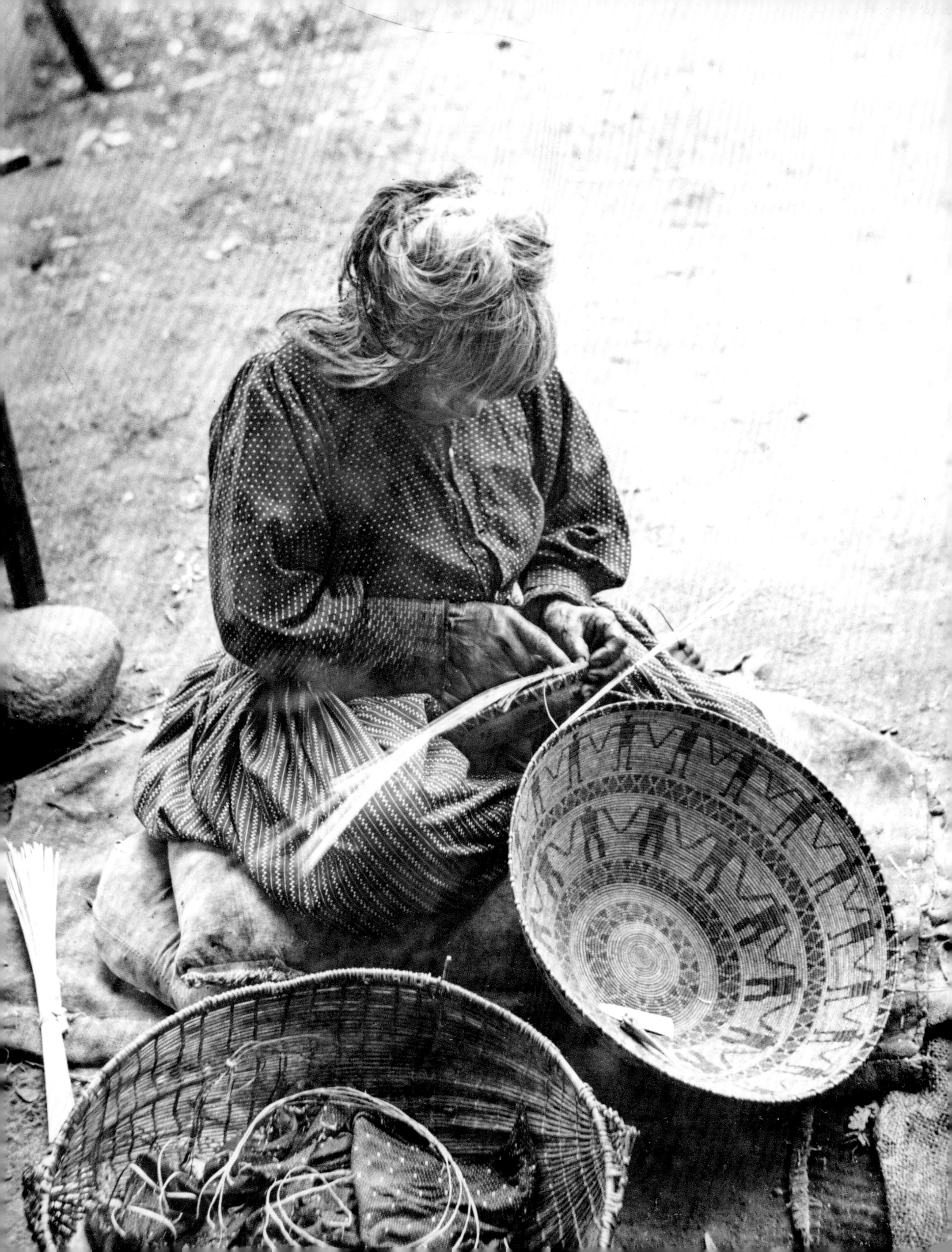

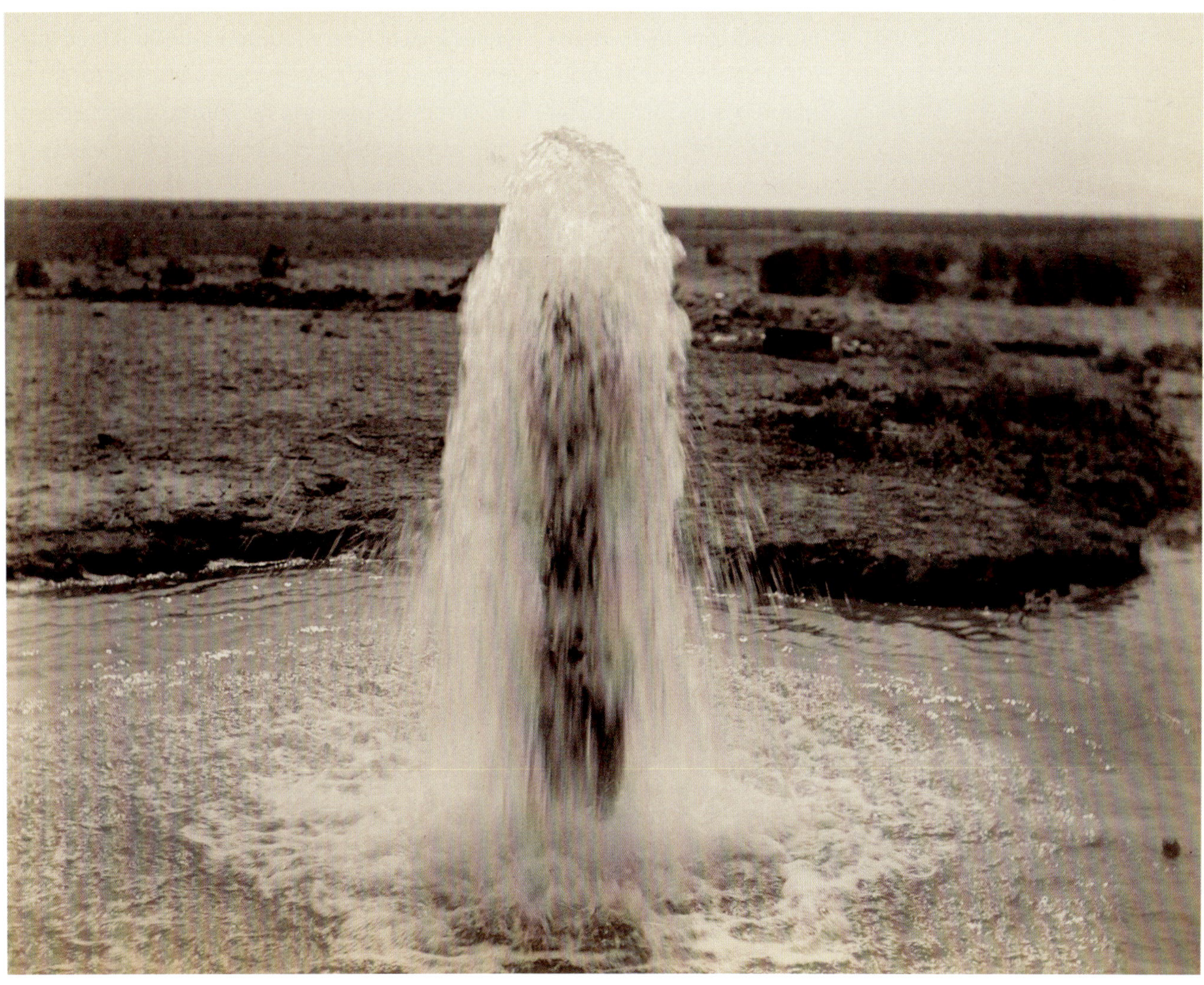

liquid profusion was only possible because of genocide carried out by U.S. settlers and military forces against the Indigenous communities of California's Central Valley. Subsequent displacement and confinement onto reservations such as the Tule River aimed to overwrite the Native practices of land use and cultivation that shaped this weaver's basket making. Yet such work continues to embody a history of horticultural practice and the judicious channeling of this landscape's water, with the baskets forming a kind of "allegorical landscape" linking past and present in continuous relation.[2]

Following the arc of this exhibition from eighteenth-century Britain to nineteenth-century California, this essay lands on three specific sites in England, Jamaica, and the United States to consider how fresh water becomes imagined as a bounded, controllable, and efficacious object in Anglo-American modernity.

Fig. 2.1. Charles C. Pierce, *Yokuts Indian Woman Weaving a Basket*, n.d. Gelatin silver print, 8 × 6 in. (20.3 × 15.2 cm). The Huntington, C. C. Pierce Collection of Photographs, photCL Pierce 02534.

Fig. 2.2. Carleton Watkins, *Artesian Well, Kern County, Cal.*, c. 1890. Gelatin silver print, 6⅝ × 8⅝ in. (17 × 22 cm). Library of Congress Prints and Photographs Division, LOT 3379.

Infrastructure, in this sense, denotes a set of interventions in the built landscape that enable or amplify water's latent capacities. Canals, aqueducts, and irrigation channels transformed water into corporate property for commodity transport; into a motive force underpinning unfree systems of labor; or into a resource for the sustenance of crops and livestock at industrial scale.[3] In that it augments the existing capacity of natural systems, infrastructure was essential in shaping a modern understanding of the natural world as the target of "improvement" through labor. Roadways, urban sanitation, railway lines, telegraphic cables: all remade the spatial order of the environment into a stage for what Lauren Berlant has called "the movement or patterning of social form," which is to say, new relations among persons, commodities, and capital that displace existing relations.[4] Infrastructure's potent symbolic and aesthetic significance—its "poetics," as Brian Larkin terms it—remains in excess of any instrumental function.[5]

In this sense, art and infrastructure are both modes of temporarily stabilizing and directing otherwise chaotic matter.[6] The act of representing freshwater infrastructure afforded an expressive means of articulating relations among art, modernity, and the environment under the sign of capitalism and imperialism. As a subject for art, infrastructure also puts pressure on how artworks are imagined as ecological systems in their own right—systems that were often shown to be leaky, ambivalent, or contradictory. In recognizing art's uncertain powers, never quite firmly aligned with the improver's view of the world, we might also temporarily destabilize the social and political histories encoded in the nineteenth century's environmental imagination.

East Anglia, England, 1816

Water in all of its forms permeates John Constable's art: as rivers moving through landscapes, as dew gathering on foliage, as clouds traversing the sky.[7] This sense of saturation, that Constable's paintings were, as one writer put it, "all humidity," would ramify in the contemporary critical response to his art.[8] Despite his delectation in seemingly formless water vapor, which he staged in numerous cloud studies (see page 115, for example), it was a rigorously formed kind of water, the navigable River Stour, that was the most enduring and substantial subject of his art. As one critic sardonically wrote in 1819, Constable's paintings "have been almost invariably composed of the same materials, i.e., Water-Mills, Water-Locks, Navigable Canals, and a Flat Cultivated Country." This, the critic felt, made for "scanty materials."[9] Constable did indeed trace and retrace the valley formed by the Stour at the border between Suffolk and Essex in southeast England in dozens of pencil drawings, oil sketches, and finished paintings. In such repetition, he found a subject through which the deep material constitution of the natural world might be plumbed and reimagined.

The River Stour is emblematic of an important history of property, profit, and ecological change in Britain: the building and extension of a network of canals and navigable rivers in the eighteenth and nineteenth centuries that underpinned the nation's unprecedented economic growth and industrialization.[10] Constable himself was formed in and by this history, given that he grew up amid the land, mills, and transportation infrastructure that his father owned along the Stour's banks.[11] These waterways were the main means of transporting goods in the era before railroads. In the case of the Stour, coal and other commodities were transported upriver from the sea while grain and bricks were brought downriver.[12] As a new means of putting goods and people into motion, the fluid "floating world" of canals and navigable rivers also suggested, as Susanna Cole has argued, how modernity unsettled the imagined spatial and social stability of British life.[13] An early, dramatic instance of these enterprises' disquieting effects could be seen in the so-called Bridgewater Canal, completed in 1761 by Francis Egerton (the third Duke of Bridgewater) to transport coal extracted from mines on his estates. One of its signal engineering feats was an aqueduct by which the canal is carried above the River Irwell. A 1793 view of this aqueduct emphasizes the almost hallucinatory nature of the scheme, in which a boat seems to float above the ground in midair, scrambling the usual order of things and the laws of gravity (fig. 2.3).[14] The Stour navigation was much less technologically complex. A parliamentary act of 1705 set out the terms of the project and the group of subscribers whose investment underpinned the work of turning the waterway into a canalized thoroughfare: dredging, shoring up the riverbanks, and, most importantly, building the pound locks and weirs that would regulate the level and flow of water.[15] Constable's father would number among the commissioners of the River Stour Navigation Company, which maintained its infrastructure in return for the collection of tolls levied on the goods transported along the waterway.[16]

Constable's 1816–17 painting *Flatford Mill ("Scene on a Navigable River")* is organized around the river's irregularly receding shape (fig. 2.4). In the foreground, a horse has been untied from the barge it was towing, led along the towpath that trails alongside the river; behind it, young boys work to propel the barge with a long pole. In the very foreground, a weathered piece of wood that seems almost to prop up the frame of the painting indicates a bridge crossing the river and arcing out of view. This is the obstacle that has caused the quiet bustle animating the scene. The horse needs to be untied so it can cross to the towpath on the other side, while the barge will pass under the bridge. Farther down the waterway, we see a wooden lock regulating the flow and level of the water and a group of buildings—the mill that is the subject of the work, and that was owned by the artist's father, Golding Constable. All of this takes place toward the very edge of the painting, crowded out by the artist's close attention to the bank of

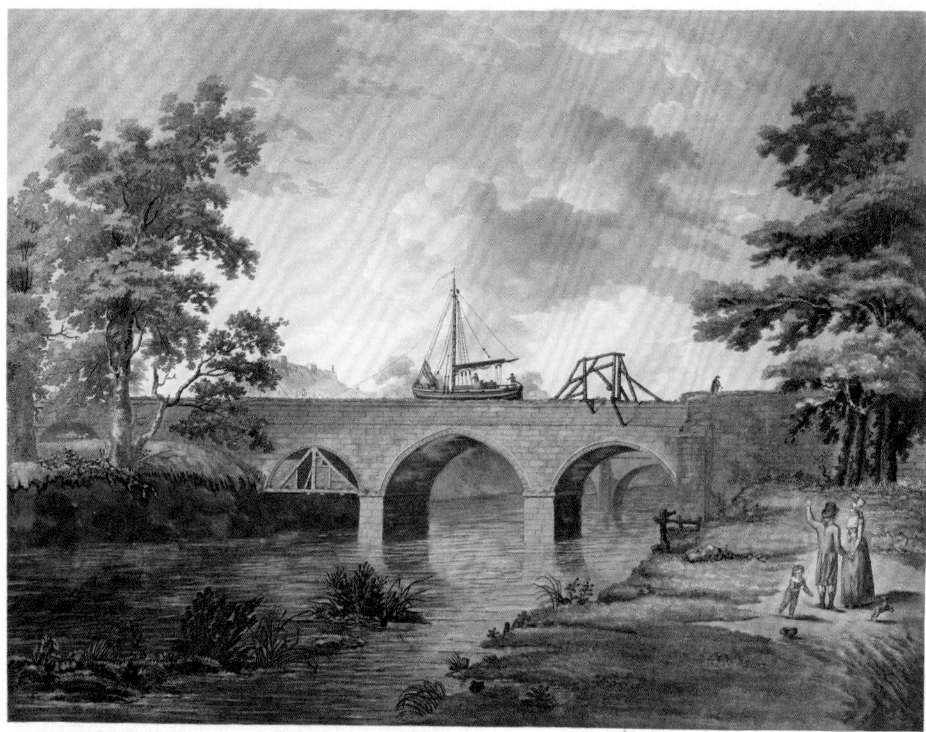

Fig. 2.3. William Orme, *View of the Celebrated Aqueduct at Barton in Lancashire*, 1793. Aquatint on paper, 17 × 22½ in. (43 × 57 cm). Science Museum / Science Museum Group, Sotheby, Parke, Bernet and Co., 1983-252.

trees running along a drainage ditch and a field stretching out beyond. Given the rough, unruly, lively world of trees, foliage, leaves, grass, sky, and air that unfolds on the right side of the painting, we might be forgiven for assuming that this patch of British landscape was more or less "natural"—that is, that Constable was painting a world that took its own course, independent of human life. This is the strong fiction that Constable's paintings aimed to embody.[17] The locks, wooden banks, and bridges of the navigable river are knitted so strongly into the order of things that the artifice they achieve—their function—has been forgotten.

What did it mean to make a canalized river look "natural"? As many art historians have argued, Constable seemed, at most points, determined to elide the artificial and regulated world of commercial production and waged labor that shaped the world of his paintings—relegating agricultural workers to near obscurity and emphasizing the seemingly autonomous world of natural things. While *Flatford Mill* suggests a complex ecological web of relations by means of

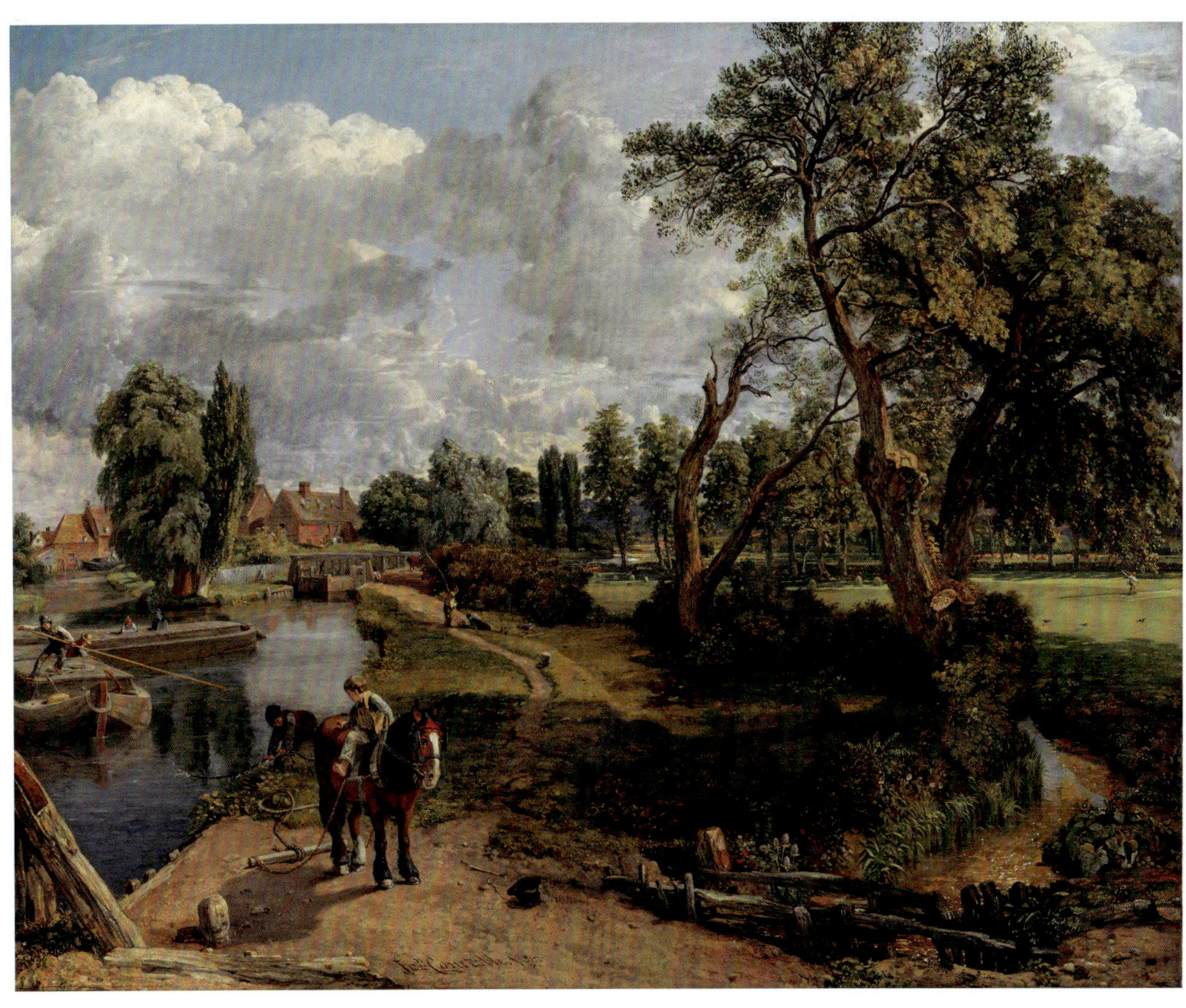

Fig. 2.4. John Constable, *Flatford Mill ("Scene on a Navigable River")*, 1816–17. Oil on canvas, 40 × 50 in. (101.2 × 127 cm). Tate, London, bequeathed by Miss Isabel Constable as the gift of Maria Louisa, Isabel, and Lionel Bicknell Constable, 1888, N01273.

a painstaking naturalism, his oil sketches, made *en plein air,* used the force and verve of brushwork to suggest the animated environmental forces (wind, moisture, sunlight) that connected the river's infrastructure with the world around it. In an earlier sketch, the lock at Flatford takes on a juicier, gestural horizontality, in which the wooden bars framing the passage of the lock echo the slashing vectors of light and cloud above (fig. 2.5). The world of the picture is startlingly mobile and uncertain, almost liquid, as elements merge into one another. In this case, the more Constable emphasized his own sensations and individuated response to nature—bodied forth by the energetic, lively brushwork of his sketches—the more he made the (increasingly volatile) cultural systems of capitalist landscape improvement, social hierarchy, and disciplined labor appear also to be "natural"—given and unchanging, echoed by and echoing the movement of elemental forces through the landscape.[18]

Yet it remains striking that Constable so intensely identified his art with the subject of a navigable river and its infrastructure, the very emblem in this moment of nature set to work.[19] His reputation was made with a series of large-scale, "six-footer" canvases depicting this landscape, which he exhibited between 1819 and 1826. As in *View on the Stour near Dedham,* this series of paintings transformed the mundane work of navigation and its surrounding world into a subject for ambitious painting (see fig. 1.7). The investment suggested in his repeated return to such subjects, the "scanty materials" of his painting, would also be echoed in his private rhetoric. In a letter of 1821, he wrote of his attachment to "the sound of water escaping from Mill dams. . . . Willows, Old rotten Banks, slimy posts, & brickwork. I love such things. . . . As long as I do paint I shall never cease to paint such Places."[20] (See, for example, the water-soaked wooden banks in the foreground of *View on the Stour.*) In this passage the act of painting is likened to the multisensory immersion in the River Stour's particular liquid-infused materiality, getting up close to the seeping and changing stuff of the landscape in which he grew up and that he would "never cease to paint." We can think of him like the boy on the barge in *View on the Stour,* leaning to press his barge pole into the muddy bank below, or the boy in *Flatford Mill* leaning over to haul in a rope dripping with water. But there is ambivalence, too. The water he imagines in his letter is "escaping" from the structure meant to contain it; the banks holding up the river's edge are rotting, and organic growth has coated them.

A visual and verbal rhetoric of disordered matter is in play here, one that is opposed to the delineation of plots of land and fixed, dredged riverbeds that would define the language of property documents and maps accumulated in the work of the River Stour Navigation Company. The question of property, and of Constable's virtual superintending of his family's lands through the medium of paint, has played a large role in the interpretation of Constable's paintings.[21] According to British common law, the riverbeds traversed by an *unnavigable*

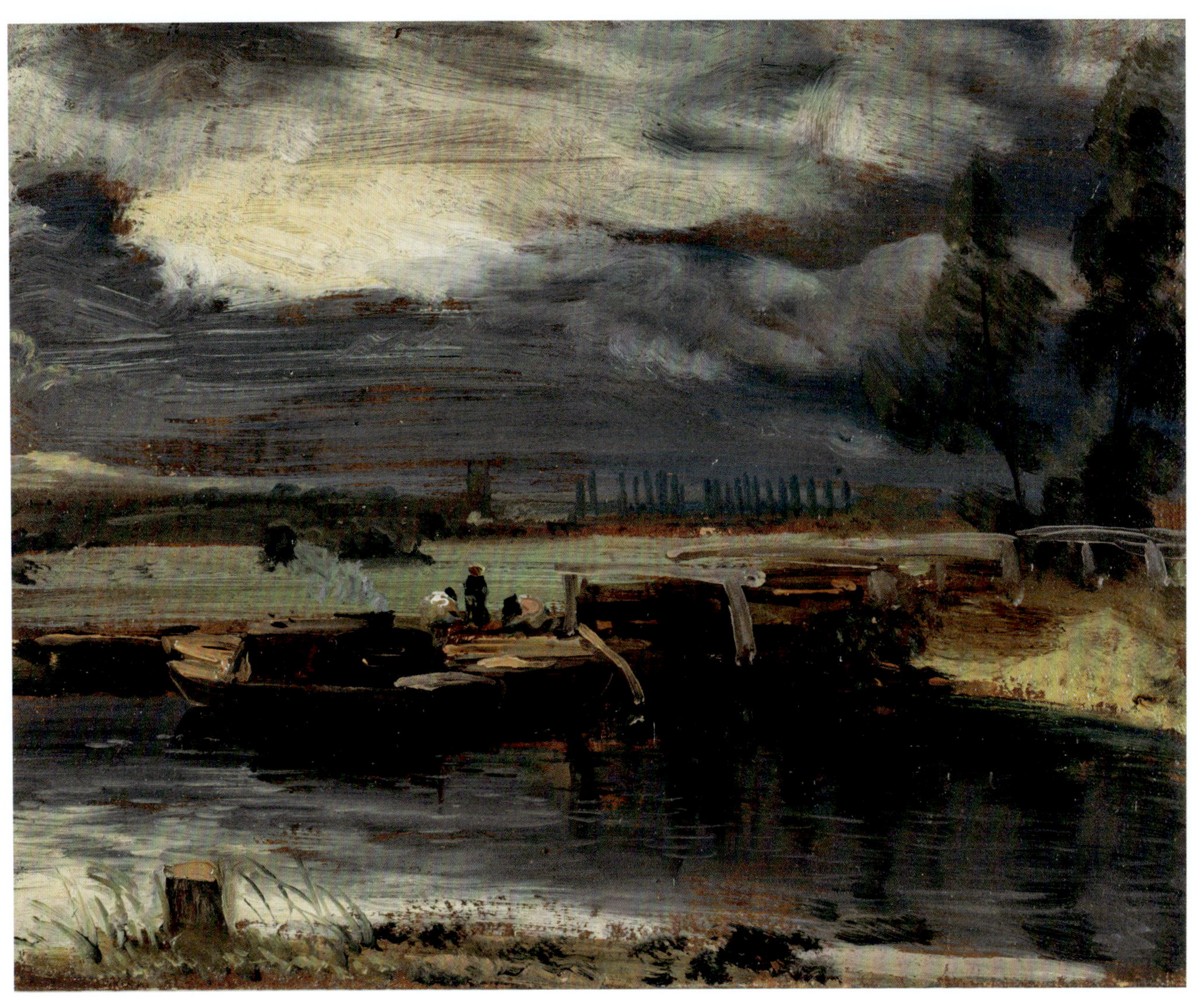

Fig. 2.5. John Constable, *Barges on the Stour, with Dedham Church in the Distance*, c. 1811. Oil on paper laid on canvas, 10¼ × 12¼ in. (26 × 31.1 cm). Victoria and Albert Museum, London, given by Isabel Constable, 1888, 325–1888.

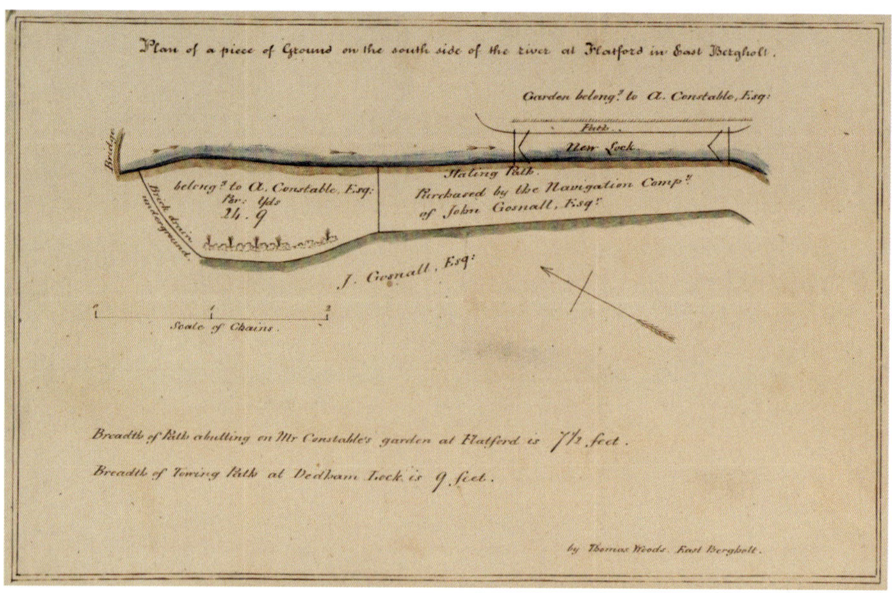

Fig. 2.6. Thomas Woods of East Bergholt, *Sudbury Borough River Stour Navigation Plan of a Piece of Ground on the South Side of the River at Flatford*, c. 1850. Suffolk Archives, Suffolk Record Office, EE501/13/5/3.

watercourse were owned by the adjacent property owners; yet if a river, such as the Stour, was navigable, it would then be deemed a "public highway." Access to navigable rivers and their resources were held in common. This sense of the river as a commons, however, was traversed by the rights of the corporate bodies that built and maintained its infrastructure, as well as the complex rights that governed the towing paths alongside rivers and canals.[22]

The spaces Constable depicted, then, were ones in which the intricate divvying-up of natural resources met in complex, tangled folds, where private interests met corporate superintendence, both of which meanwhile crowded out common access. In one undated map that records the purchase of land by the Navigation Company, we see how minutely and precisely the land around Flatford Mill has been divided by owners and functions (fig. 2.6). The tract near the bridge over the Stour (where Constable stood to paint *Flatford Mill*) is shown to be the property of the Constables; the quasi-common margin of the towing path traversing different plots is traced in tan ink.[23] We might see the figures in *Flatford Mill* as dancing around the various, invisibly inscribed jurisdictions of ownership and rights, a delicacy of movement governed by property that is posed against the straining, expansive forms of the trees to the right of the picture. Such smaller-scale property relations and rights were set into the larger geography traced in an 1822 map of the lower River Stour (referred to here as "Stower," an alternate spelling; fig. 2.7). It describes a watery landscape made over into infrastructure, one in which coal and grain might move seamlessly along lines traced across the terrain in a choreographed movement. This was a language of

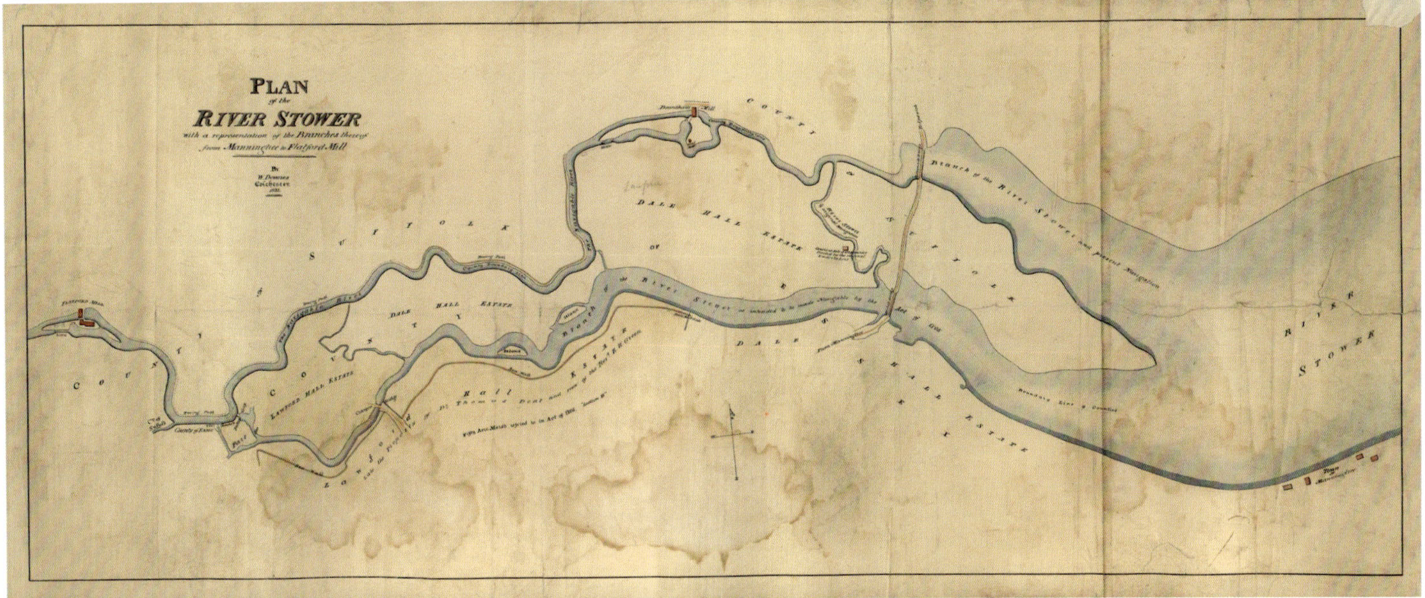

Fig. 2.7. W. Downes of Colchester, *Sudbury Borough River Stour Navigation Plan of the River Stour*, 1822. Suffolk Archives, Suffolk Record Office, EE501/13/5/2.

water drawn in lines and arrows indicating the stable flow of commodities in both directions.

Despite the imagined seamlessness of circulation described in the map, Constable was more interested in moments of friction: in staging energetic, if allusive, actions where circulation around the river's navigations becomes complicated or momentarily thwarted, as in the tangled movements of horse and humans in *Flatford Mill*. In *The White Horse*, we are presented with the strange spectacle of a barge-towing horse being ferried across the river at a spot where the towpath switches from one side to the other (fig. 2.8); in *The Leaping Horse*, the tow horse now needs to jump over the bounds of a barrier erected to control the movement of cattle, turning the mundane work of drawing barges along the river into a strangely heroic, condensed moment of action (fig. 2.9). Such gestural charge is concentrated in the human figure in *The Lock*, in which we see a man opening one of the navigation's pound locks—the key development in inland navigation, which controlled the level of water in the chambers (fig. 2.10).[24] The lock itself had been an object of concern for some years; the artist's brother Abram wrote to the Navigation Company's commissioners in 1820 to complain that the locks were "in a ruinous state and in great decay."[25] The 1825 painting seems to show it instead in a more intermediate state—still functioning, yet not quite separate from the creeping vegetation that surrounds it. Attention focuses on two gestures of force: that of the red-shirted worker manipulating the lock and that of the other bargeman using his body's weight upon a rope to keep the barge steady. Below the lock, Constable's thick white impasto suggests the churning

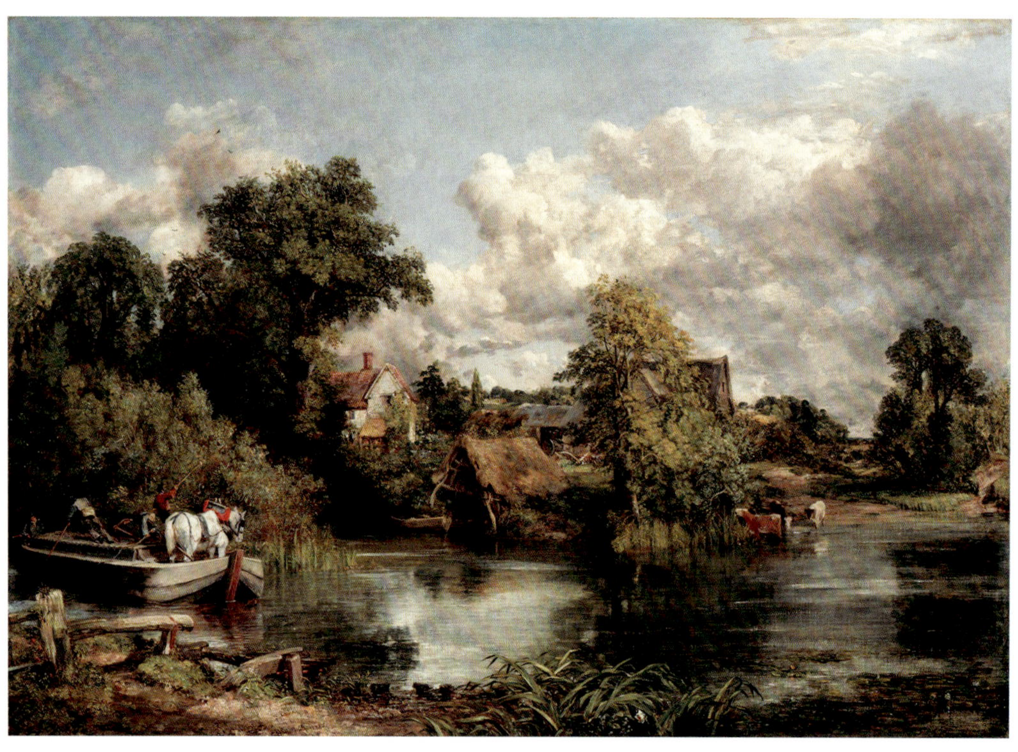

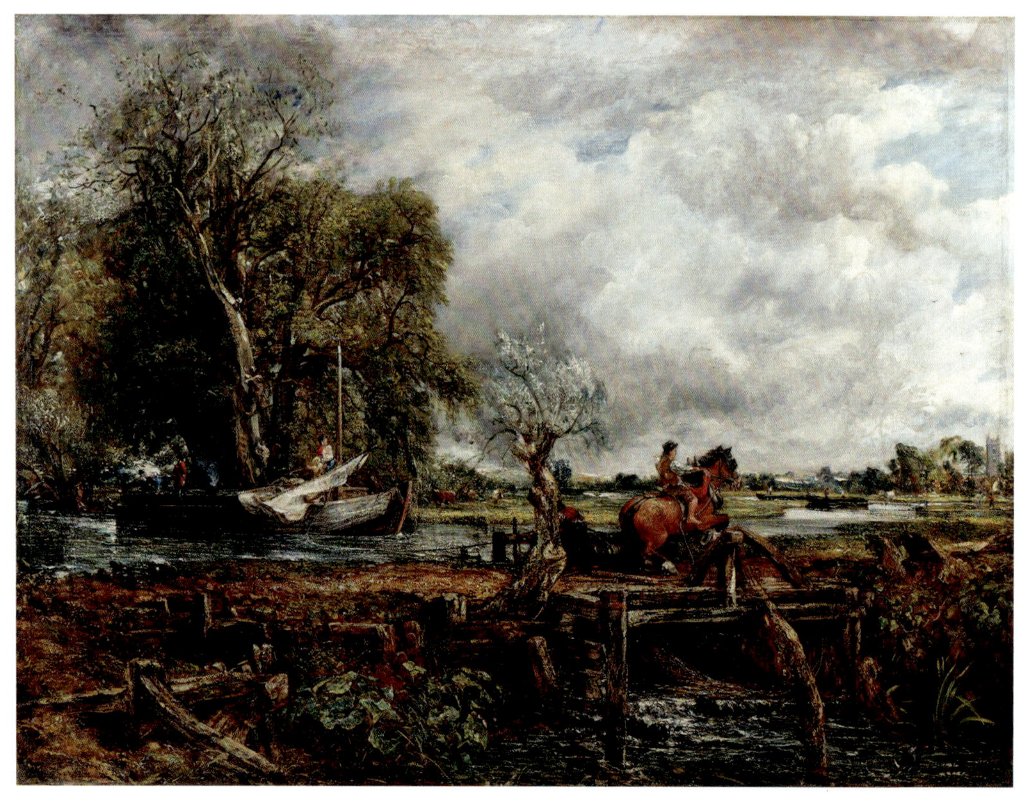

Fig. 2.8. John Constable, *The White Horse*, 1819. Oil on canvas, 51¾ × 74⅛ in. (131.4 × 188.3 cm). The Frick Collection, New York, purchased by The Frick Collection, 1943, 1943.1.147.

Fig. 2.9. John Constable, *The Leaping Horse*, 1825. Oil on canvas, 55⅞ × 73¾ in. (142 × 187.3 cm). Royal Academy of Arts, London, given by Mrs. Dawkins, 1889, 03/1391, PL027570.

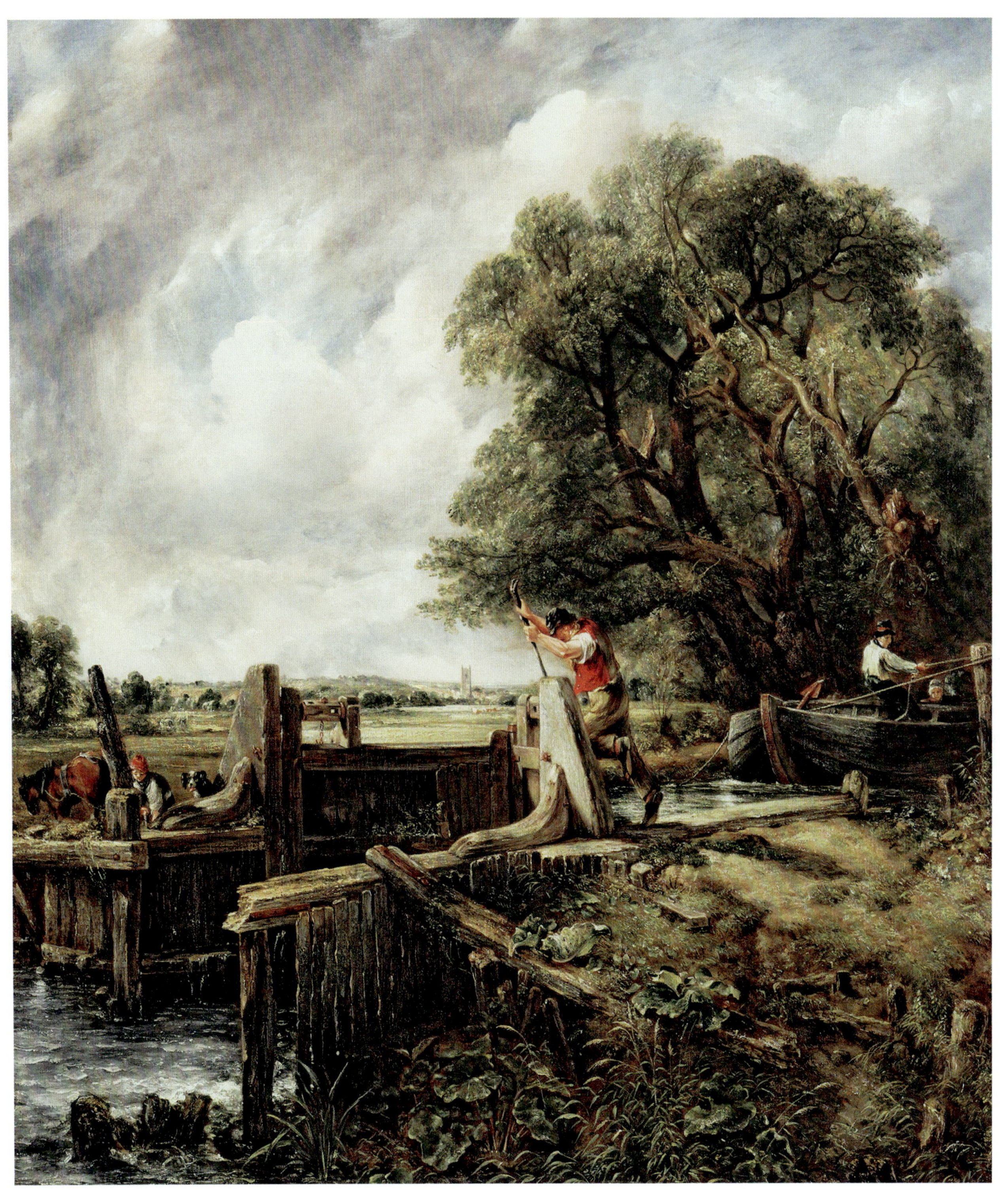

Fig. 2.10. John Constable, *The Lock*, 1824.
Oil on canvas, 56 × 47½ in. (142.2 × 120.7 cm).
Private Collection/Bridgeman Images.

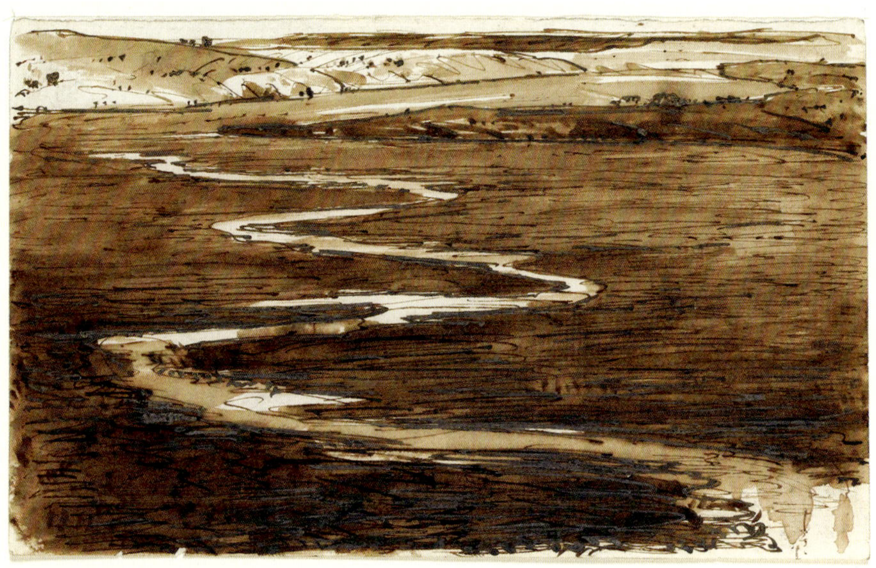

Fig. 2.11. John Constable, *Stour Valley*, 1830. Pen, brown ink, and wash on paper, 4⅜ × 7¼ in. (11.2 × 18.4 cm). The Fitzwilliam Museum, Cambridge, given by John Charrington, 1910, 3314b.

and disturbance of the water as it flows out. The central figure in *The Lock* is thus poised at a hinge between two conceptions of the natural world that Constable sought to hold in a tenuous balance—the landscape that requires "improvement" such that the river can be put to work (via the labor of animals and humans), and the landscape that exerts its own animacy and force as it undermines, corrodes, and "seeps" through its contained banks.[26]

We know that Constable paid deep attention to how water shaped the landscape, in ways perhaps less dramatic than the expressive gesture of his red-shirted worker. The forces staged in *The Lock* might be posed against that imagined in an ink drawing from around 1830, drawn for the engraver David Lucas, who was at work on a series of mezzotint engravings after the artist's work (fig. 2.11).[27] Here, Constable produces a kind of diagram of the Stour's riverine action, attempting to describe for Lucas how the movement of water along the winding stream (traced in ink) might slowly carve out the wide, fertile valley around it by means of erosion and a slowly shifting course. The scribbles of ink in the flat plains surrounding the river suggest the traces of some past action of water on the landscape, while the river itself—showing forth at points as the white reserve of the paper—seems to etch its course through the layers of ink and wash on the sheet. In articulating to Lucas the desire to understand the deeper causes of this landscape's constitution, Constable recognized the Stour as the primary shaper of the world he depicted—perhaps not an "agent," but certainly a force with and against which human life took shape. To this balance he added the work of painting, a mode of material manipulation in which both forces—the orthogonal, linear world of human construction and the seeping, dispersive world of natural force—might find parallel, if unresolved, concord with one another. (It would likely not have escaped Constable's attention that an artist, Leonardo da Vinci, had invented the pound lock, though he likely would not have known da Vinci's wild drawings depicting water's swirling force.)[28] In this way, the laborer opening the lock is a strange emblem of the work of the painter, held in the balance between portraying the natural world as it acts upon the landscape and repetitively reshaping that world to serve human ends.

St. Mary Parish, Jamaica, 1821–22

The visual imagination of the industrial revolution—and its effects upon landscapes and environments—has often centered on the image of the coal- and steam-powered factory. Yet water served as the first motive power behind the remaking of the world under industrial capitalism, just as canals and navigable rivers transformed the circulation of goods before the era of the railroad. Despite advances in steam power in the late eighteenth century, the transition from water-powered to steam-powered mills and factories only occurred later, in the 1830s.[29] This same imagination of industrialization often also excludes the other central sites of production: plantations in the Americas where sugar, cotton, indigo, and other so-called raw materials were produced under regimes of enslaved labor. Jamaica was, for much of the eighteenth and early nineteenth centuries, the richest and most important British colony in the Caribbean, and a site for the accumulation and dispersal of massive wealth. In the plantation, which Sidney Mintz would term the "synthesis of field and factory," the ideologies of "improvement" took on a massive scale and a new violence.[30]

In their encounters with the Jamaican landscape and environment, British writers and artists characterized it both as a luxuriant, beautiful landscape and as a space of extremity and dangerous alterity—depending on their ideological bent and purpose.[31] As many have shown, the picturesque, a mode of viewing, describing, and designing landscapes so they would resolve into pleasing formats, emerged as a kind of visual infrastructure that could transform the Jamaican landscape into a familiar aesthetic object. In other views of Jamaica, like the engravings made after painter George Robertson's views of Jamaican plantations, the central structuring fact of chattel slavery was disguised by the landscape's refashioning to fit the conventions for representing English rural scenery and its pastoral appearance (fig. 2.12).[32] The rushing Roaring River in the engraving's foreground suggests a sensory immediacy, a natural force, that displaces attention. Such aesthetic refashioning mirrored the physical transformation of the landscape to serve the large-scale monoculture of the sugar plantation, which required a vast economy of scale and an expansive technological infrastructure—especially for irrigation and water power—in order to be profitable.

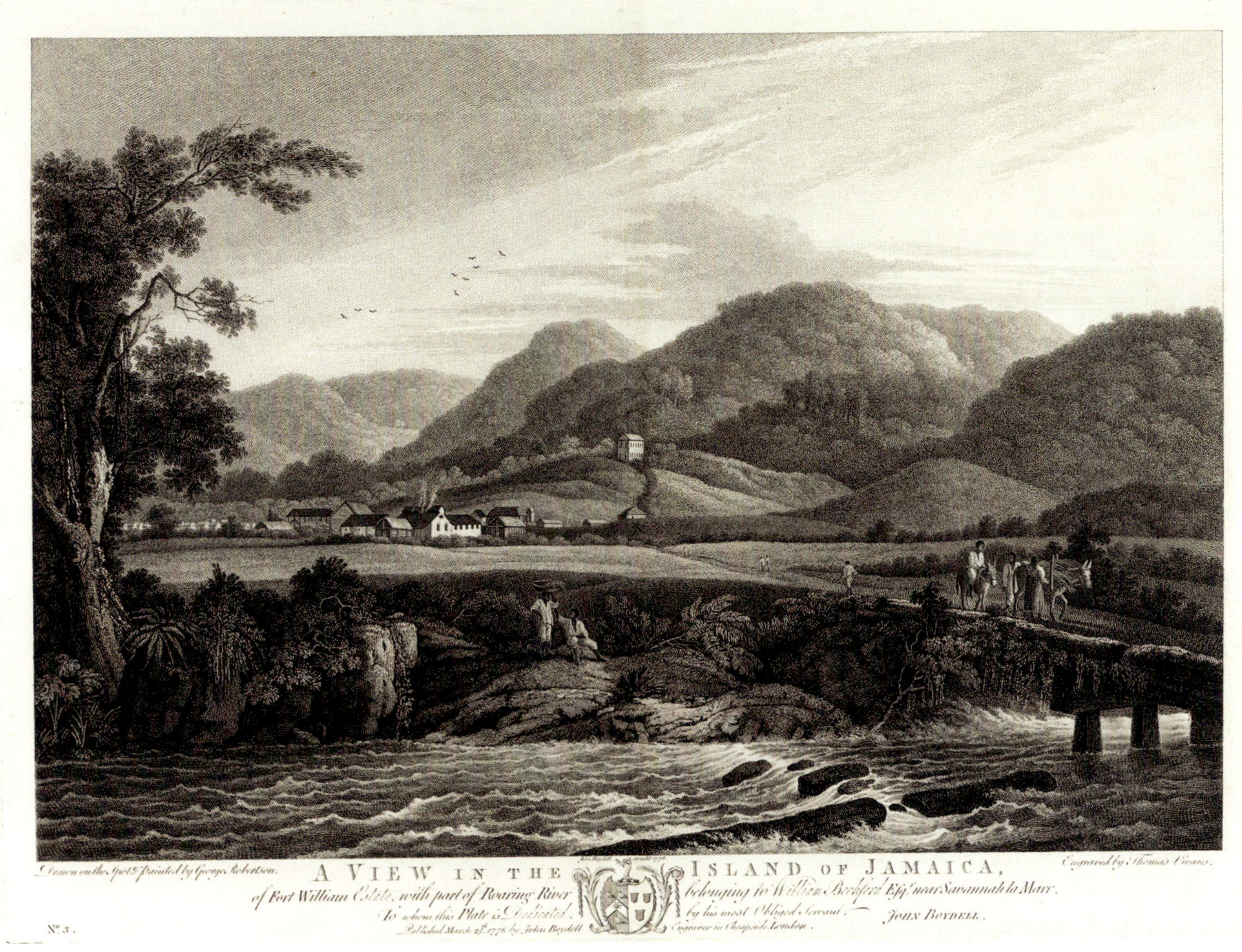

Fig. 2.12. Thomas Vivares, *A View in the Island of Jamaica, of Fort William Estate, with Part of Roaring River Belonging to Mr. William Beckford, Esq.r near Savannah la Marr*, 1778. Engraving, sheet: 18½ × 22$^{5}/_{16}$ in. (47 × 56.7 cm). Courtesy of Yale Center for British Art, Paul Mellon Collection, B2001.2.1520.

The picturesque and the infrastructural converge in a hand-colored aquatint view of Trinity Estate, one of a series of neighboring estates owned by the Bayly family in St. Mary Parish, Jamaica (fig. 2.13). Published in 1824, this aquatint formed part of the series *A Picturesque Tour of the Island of Jamaica*, produced by the architect and artist James Hakewill. This series aimed to present an image of the Jamaican sugar plantation in which the violent realities of work and subjugation yielded to the depiction of a natural world that appeared to generate the plantation's produce of its own accord.[33] In the accompanying letterpress, Hakewill describes Trinity Estate as "one of the most desirable properties in the Island," due to the "richness of the land," the "general healthiness of the spot," and the "excellent provision grounds" for enslaved workers.[34] Though the elements fit together somewhat awkwardly, Hakewill's image conforms to the generic constraints of picturesque views: a winding path leading our eye into the image, foliage framing the scene, a river that reflects the placid sky above it. The reflecting capacity of water, like the river that curves through Hakewill's view, was essential for the transforming act of aesthetic appreciation that converted plantations into

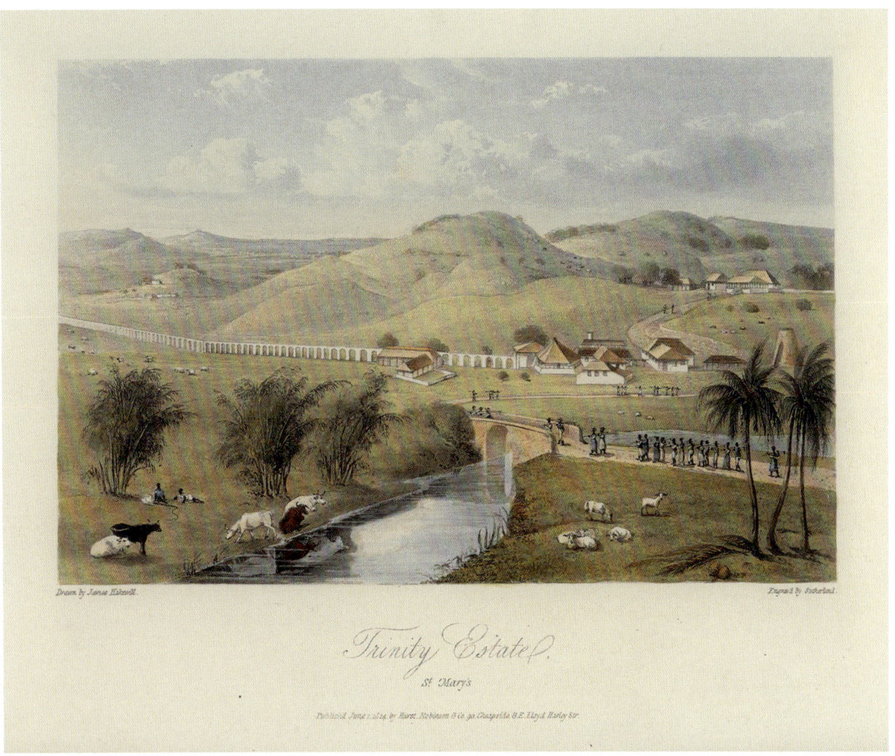

pictures. "Those [estates] that abound with water," wrote the enslaver and writer William Beckford, "present the most pleasing variety of scenes; for without water, let the view be ever so extensive, the landscape cannot be said to be really perfect."[35]

Fig. 2.13. "Trinity Estate, St. Mary's" by Thomas Sutherland (engraver) after James Hakewill in *A Picturesque Tour of the Island of Jamaica, from Drawings Made in the Years 1820 and 1821*, no. 4 (London: Hurst and Robinson, 1824). Courtesy of Yale Center for British Art, Paul Mellon Collection, T 683 (Fo. A).

Unlike many of Hakewill's views, *Trinity Estate* also features a group of enslaved workers traveling along the path in an orderly row, their work forming a contrast to the "picturesque and pleasing group" (to use Beckford's term) of figures resting under a stalk of foliage. This false image of bondage would repeat throughout representations of New World slavery.[36] Yet an equal weight is given in Hakewill's image to the stone aqueduct that extends from the left edge of the image toward the works of the plantation, centered on the water mill. On Jamaican plantations, water was commonly considered the ideal source of power to run the mills that ground sugarcane into juice that would, in turn, be distilled and refined. These assemblages built to channel water's force were the most complex and expensive element of environmental infrastructure.[37] Unlike windmills or cattle-powered mills, water mills could be depended upon to provide more constant motive power, although sugarcane harvests still would often coincide with a period of water scarcity that could limit the mills' function.[38] In this sense, two imperial fantasies about the labor that powered the plantation are posed against one another.

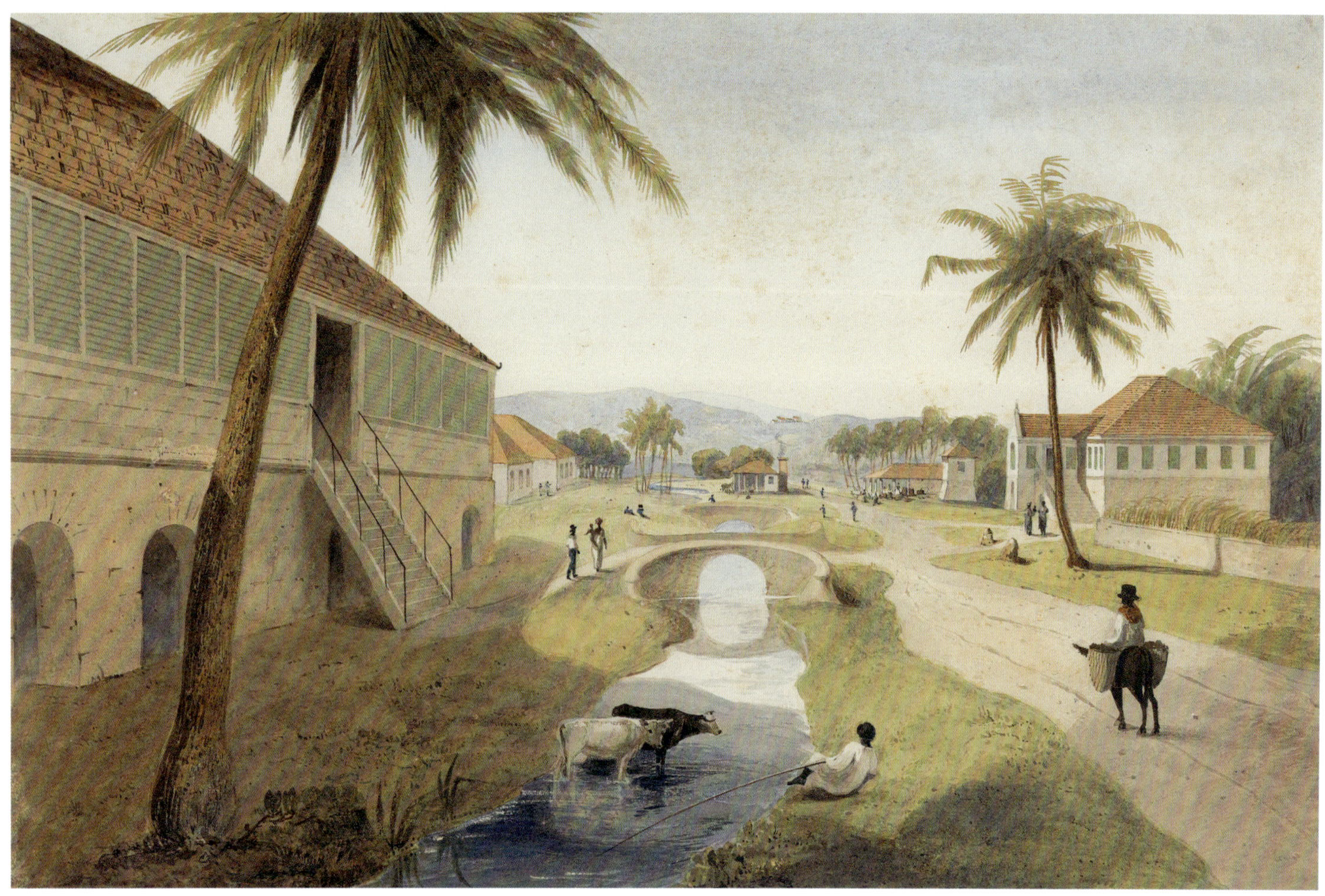

Fig. 2.14. James Hakewill, *Holland Estate, St. Thomas in the East, Jamaica*, c. 1820–21. Watercolor on moderately thick, slightly textured, cream wove paper, sheet: 12 × 16½ in. (30.5 × 41.9 cm). Courtesy of Yale Center for British Art, Paul Mellon Collection, B1977.14.1964.

Trinity Estate's aqueduct is "at once," Hakewill writes, "an object of utility and ornament," having been completed in 1797 at "vast expense."[39] And in his other watercolor and aquatint views of Jamaican plantations, well-ordered waterways and other aqueducts served as integral means of structuring his views (figs. 2.14, 2.15). As "ornament," these built waterways lend a sense of order and symmetry to a landscape that might otherwise be marked off as formless and strange. Shown from a great distance, as if the artist was hovering over the scene, Trinity Estate appears to be a self-contained, functioning complex, operating as if without human labor. As the Martiniquais poet Édouard Glissant limns, the Caribbean plantation was supposed to be an "enclosed place," a "closed circle"—"functioning apparently as an autarky but actually dependent" on what Glissant calls "someplace elsewhere."[40] Although the picturesque structure of Hakewill's image attempts to produce Trinity Estate as a world apart, this "dependence" creeps in: for one, by means of the implicit and massive (usually British) capital required to build and maintain the plantation infrastructure, signaled by the aqueduct. Its scale and extension beyond the image's frame implicates the contingencies of ecology and climate upon which the plantation system depended.

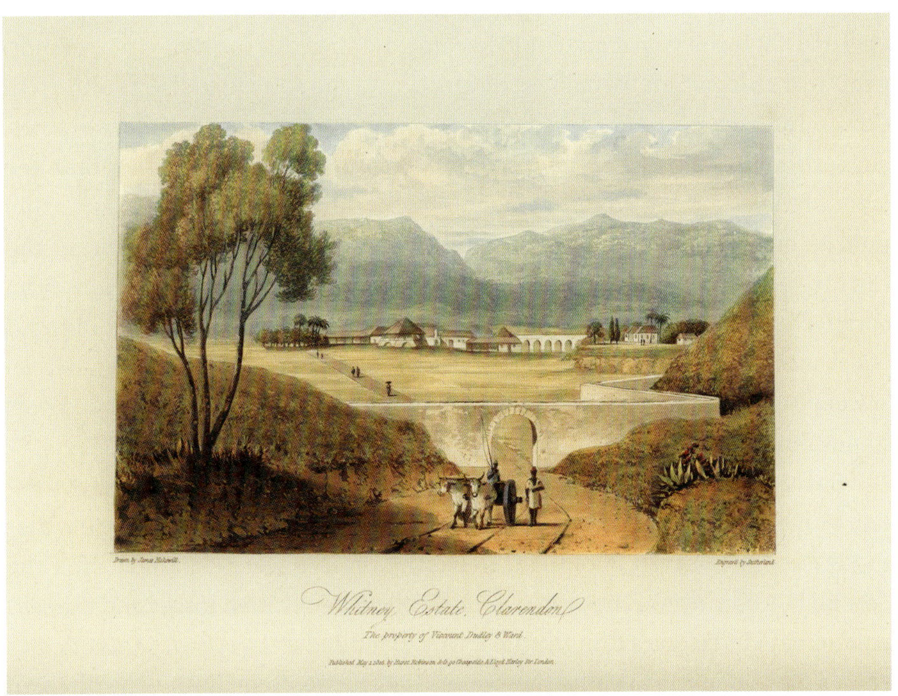

Fig. 2.15. "Whitney Estate, Clarendon" by Thomas Sutherland (engraver) after James Hakewill in *A Picturesque Tour of the Island of Jamaica, from Drawings Made in the Years 1820 and 1821*, no. 6 (London: Hurst and Robinson, 1825). Courtesy of Yale Center for British Art, Paul Mellon Collection, T 683 (Fo. A).

Without large areas of forested land to anchor the view, the edges of Hakewill's image are uncertain. Indeed, the image testifies to the extent of the systematic devastation of Jamaican ecologies under plantation monoculture. While the intensive forest clearance was imagined as making Jamaica "healthier" for White settlers, it also became clear to European commentators by the eighteenth century that the soil depletion and erosion caused by this massive terraforming was rapidly degrading the environments of many Caribbean islands.[41]

Trinity Estate's aqueduct and its channeling of motive power becomes an emblem for the plantation's larger social and economic structure. In his view, the form of the aqueduct and the form of the enslaved workers walking on the road are made disturbingly equivalent. Distributed into two repetitive, static rows, linked together by bundles perched on their shoulders, the workers seem to be depicted as themselves a kind of machine of production akin to the water infrastructure powering the mill. Such dehumanization was consonant with the ways that colonial discourse stripped enslaved workers of their agency, instead displacing vivacity and vitality upon nonliving systems of commerce and circulation. In his 1774 *History of Jamaica*, Edward Long described the recent improvement of the road system in Jamaica: "it seems as if the whole island had been suddenly animated," as if these new roads provided a "vital principle which has infused all these symptoms of vigour, agility, and health, into the whole mass, and rouzed it into active life."[42] In a world order that turned human lives into objects, by converse logic the landscape could become the stage for the animation of inorganic technology. This "ornamental" function of the aqueduct directs our attention to infrastructure's ordered extraction of power from natural resources, suggesting these works as the locus of the plantation's profit (rather than the labor of the enslaved). Here we might see how infrastructures, as Brian Larkin puts it, "emerge out of and store within them forms of desire and fantasy and can take on fetish-like aspects that sometimes can be wholly autonomous from their technical function."[43] The aqueduct powered the plantation's machines, but it also was an aesthetic object that was symbolic of an imperial and commercial system that seemed to possess its own quasi-organic life and aimed to disavow its devaluation and theft of human labor.

But the order posited by the aqueduct's neat rows, and the orderly circulation of energy such infrastructure suggested, had other aims. As one writer in 1823 noted, the system of roads that Long praised were expanded after the Second Maroon War of the 1790s, a conflict between British forces and the self-emancipated Maroon communities; this extended network was intended to enable British militia to violently suppress any further rebellion.[44] The threat of revolt would have been particularly urgent to address for an artist picturing Trinity Estate, which was well known as the site where many of the Black insurgents in Tacky's Rebellion, a major and widespread revolt against the plantation class in 1760–61, lived and worked. In his letterpress, Hakewill quotes passages from the *History* written by Bryan Edwards, nephew of the plantation's then-owner Zachary Bayly; that same book contained a detailed, inflammatory account of Tacky's Rebellion as it had unfolded from Trinity Estate.[45] Perhaps this is why this plate from Hakewill's series so insistently renders its human subjects static and rigidly ordered, while emphasizing the capacity of the plantation system to transform the landscape.

We might then see in this print the fundamental ambivalence embedded in British understandings of imperial landscapes, and the plantation in particular. They were seen as "fundamentally alien places, atavistic spaces of degeneracy and violence . . . on the margins of European progress," as Vincent Brown writes, and yet this view, as he argues, belies the absolutely intrinsic role of plantation slavery in the industrial revolution and the very notions of "European progress" embodied by understandings of British industrialization.[46] Looking to the ways in which plantation landscapes and their infrastructure were represented is not to shift attention away from the plantation as a site of subjugation and resistance.[47] Rather, it extends a consideration of how images of nature served to mediate imperial understandings of colonial space and history, enacting a doubled stabilization of the landscape by means of infrastructure and by means of aesthetic form.

The static and placid terrain Hakewill describes in 1825 belied the turmoil

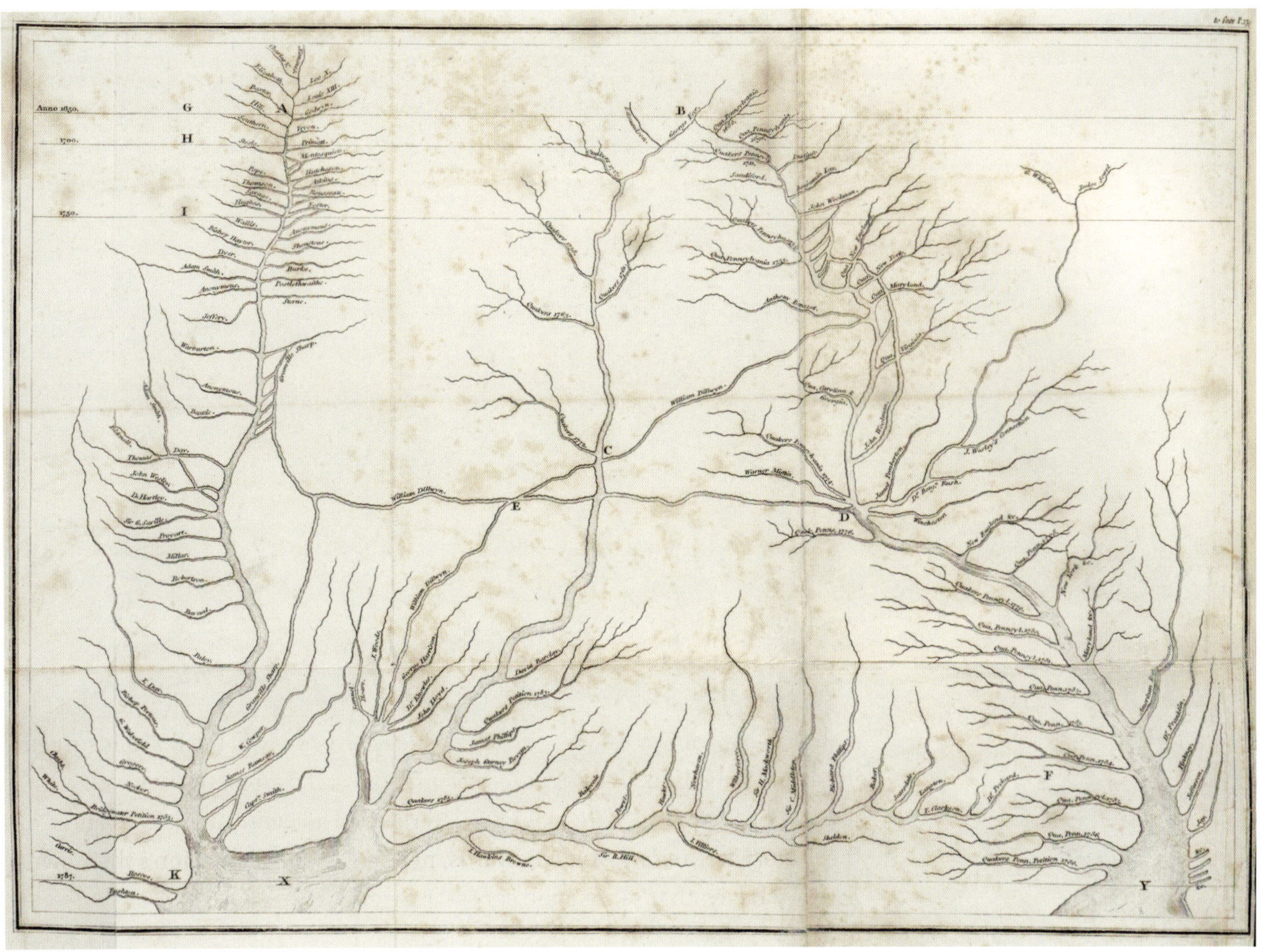

Fig. 2.16. "Untitled (Map of Abolition of the Slave Trade)," engraving in Thomas Clarkson's *History of the Rise, Progress, and Accomplishment of the Abolition of the African Slave-Trade by the British Parliament* (London: Longman, Hurst, Rees, and Orme, 1808). Courtesy of Yale University Library, Beinecke Rare Book and Manuscript Library.

that shaped the Atlantic world in the wake of continuous revolt, acts that were echoed, in other forms, by movements to abolish the slave trade in Britain. Out of the visual culture of the British abolitionist movement, a different but related slippage between natural and human histories would emerge in the widely circulated "map" published in Thomas Clarkson's *History of the Rise, Progress, and Accomplishment of the Abolition of the African Slave-Trade by the British Parliament* (fig. 2.16). It depicted the various White British and North American writers and members of the abolitionist movement as the convergence of "many springs or rivulets." These waterways "trac[e] the different streams," as Clarkson writes, "from whence the torrent arose, which has now happily swept away the Slave-trade."[48] The image uses these waterways to impart an ambivalent sense of agency to the history of abolition; while the volume celebrates the individual achievement of White abolitionists and displaces the decisive role of Black revolt, it also transforms political and social change into a landscape of inevitability. Human

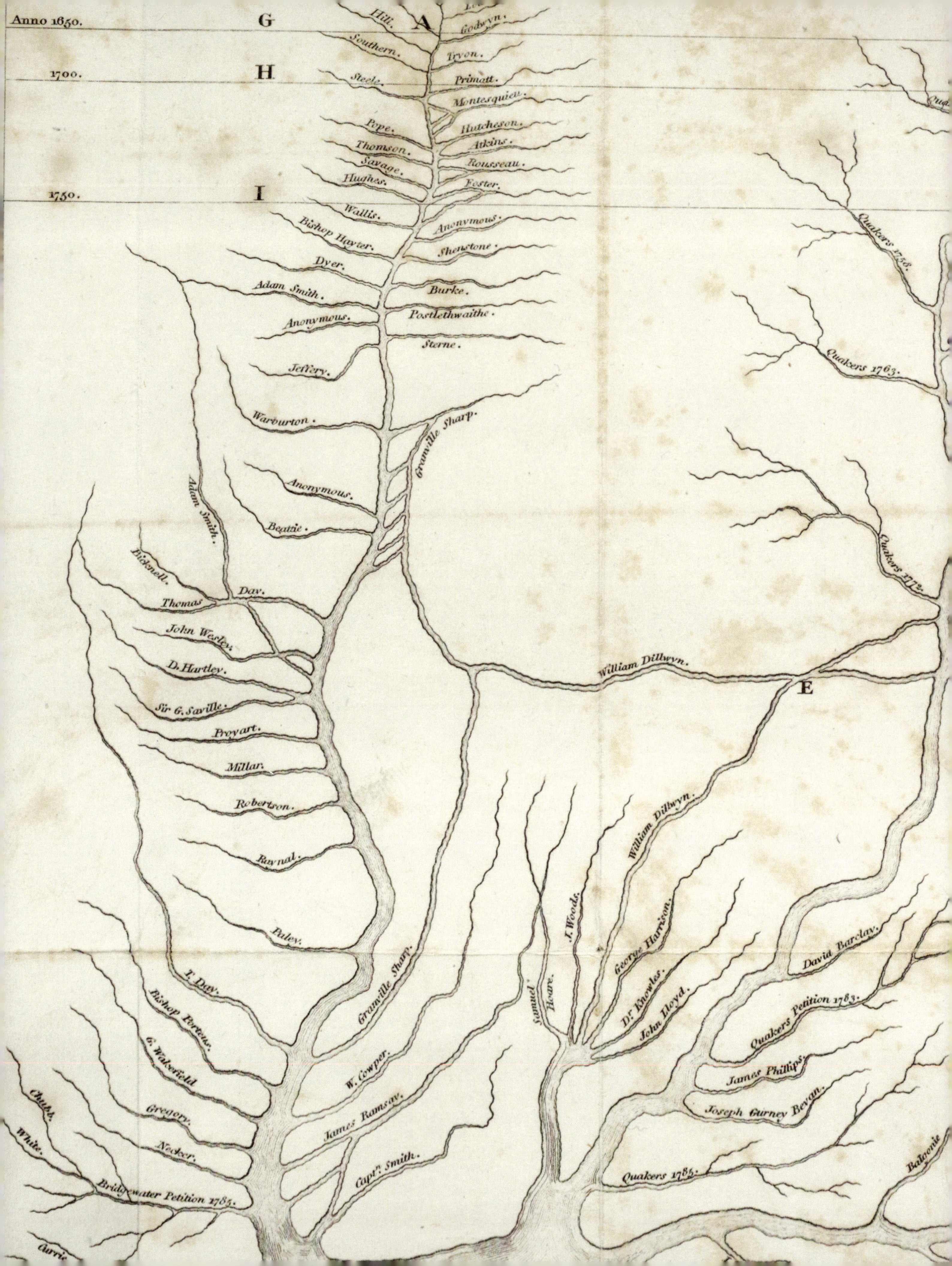
Anno 1650.
G
A
Hill.
Godwyn.
Southern.
Tryon.
1700.
H
Steele.
Primatt.
Montesquieu.
Pope.
Hutcheson.
Thomson.
Atkins.
Savage.
Rousseau.
Hughes.
Foster.
1750.
I
Wallis.
Anonymous.
Bishop Hayter.
Shenstone.
Dyer.
Adam Smith.
Burke.
Anonymous.
Postlethwaithe.
Sterne.
Jeffery.
Warburton.
Granville Sharp.
Anonymous.
Beattie.
Adam Smith.
Bicknell.
Thomas Day.
John Wesley.
D. Hartley.
Sir G. Saville.
Provart.
Millar.
Robertson.
Raynal.
Paley.
T. Day.
Bishop Porteus.
G. Wakefield.
Gregory.
Necker.
Chubb.
White.
Bridgwater Petition 1785.
Currie
Granville Sharp.
W. Cowper.
James Ramsay.
Captn. Smith.
Samuel Hoare.
J. Woods.
William Dillwyn.
George Harrison.
Dr. Knowles.
John Lloyd.
William Dillwyn.
E
Quakers 1758.
Quakers 1763.
Quakers 1772.
David Barclay.
Quakers Petition 1783.
James Phillips.
Joseph Gurney Bevan.
Quakers 1785.
Balgonie

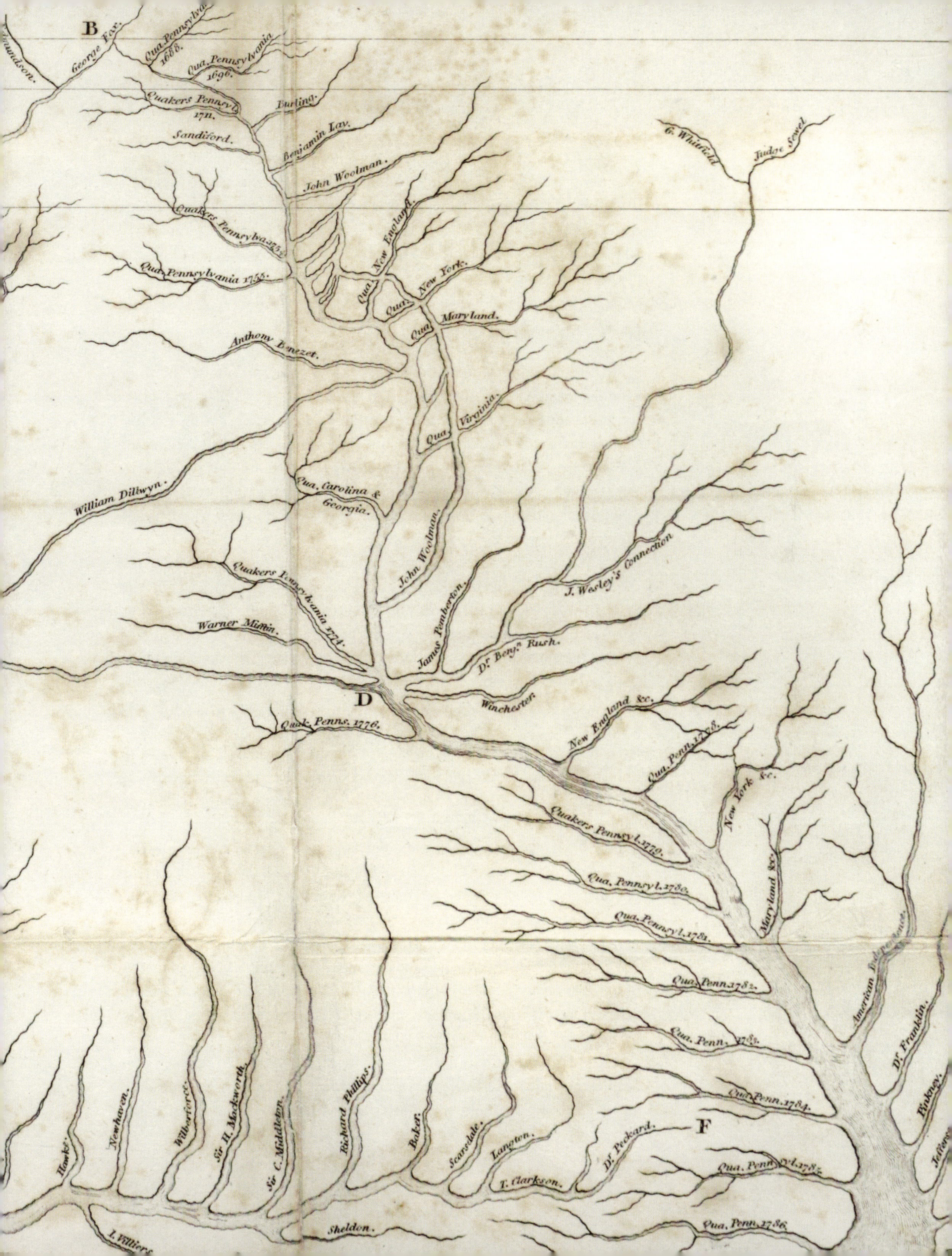
B
George Fox.
Qua. Pennsylva 1688.
Qua. Pennsylvania 1696.
Quakers Pennsylva 1711.
Burling.
Sandiford.
Benjamin Lay.
John Woolman.
Quakers Pennsylva 1754.
Qua. Pennsylvania 1755.
Qua. New England.
Qua. New York.
Qua. Maryland.
Anthony Benezet.
Qua. Virginia.
William Dillwyn.
Qua. Carolina & Georgia.
John Woolman.
Quakers Pennsylvania 1774.
Warner Mifflin.
James Pemberton.
Dr. Benjn. Rush.
J. Wesley's Connection
G. Whitefield
Judge Sewel
D
Winchester.
Quak. Penns. 1776.
New England &c.
Qua. Penn. 1778.
New York &c.
Quakers Pennsyl. 1779.
Qua. Pennsyl. 1780.
Qua. Pennsyl. 1781.
Maryland &c.
Qua. Penn. 1782.
American Independence.
Qua. Penn. 1783.
Dr. Franklin.
Qua. Penn. 1784.
Pinkney.
F
Dr. Peckard.
Qua. Pennsyl. 1785.
Qua. Penn. 1786.
Hawke.
Newhaven.
Wilberforce.
Sir H. Mackworth.
Sir C. Middleton.
Richard Phillips.
Baker.
Scarsdale.
Langton.
T. Clarkson.
Sheldon.
I. Villiers

history is framed as a kind of natural occurrence rather than a terrain of struggle and contestation.[49] And so if Hakewill would present the flow of water as a natural emblem for the plantation's power—for its capacity for self-perpetuation and regulation—Clarkson's map suggests how images of water and its capacity to shape space and history might serve to stabilize the otherwise fractious histories shaping the Atlantic world.

Kern County, California, 1881

In 1881, the photographer Carleton Watkins arrived in Kern County, California, in the San Joaquin Valley, at the request of his clients, Henry Miller and Charles Lux. Miller and Lux were cattle ranchers on an industrial scale, having acquired massive tracts of land in the San Joaquin Valley in the decades following the Mexican-American War. This land—relatively arid, though also marked by marshes and seasonal waterways—required intensive irrigation for use as agricultural and ranching land. Watkins's purpose was highly instrumental: locked in a legal dispute over water rights in the region, Miller and Lux's attorney wanted Watkins to produce photographs that might serve as evidence in the ensuing lawsuit.[50] Both Lux and Miller and the neighboring industrial rancher, James Ben Ali Haggin, had consolidated vast land holdings whose value depended upon their access to water for irrigating crops. Both had constructed extensive networks of irrigation canals to secure that access, but Lux and Miller felt that Haggin's enormous canals infringed on their rights by drawing a disproportionate flow from the Kern River. In this assertion they were supported by English common law, which argued that all property owners along a watercourse had equal rights to use the water. Haggin, however, was an "appropriationist," arguing that any person with access to water had the right to divert it for their own use without restriction. This contest over property in land and resources and a sense of their equitable or proper division was contingent upon the prior erasure of Indigenous sovereignty and the genocidal intent to destroy the lifeways of Yokuts and Chumash people in the San Joaquin Valley and the Tübatulabal and Owens Valley Paiute in the Kern River Valley. Pierce's photograph of the weaving woman at work suggests how this violent project of erasure did not succeed (see fig. 2.1).[51] But Anglo-colonial understandings of resources and their division were enshrined and enforced in law. And this particular case, *Lux v. Haggin*, would come to define the tangled mesh of water rights in the arid West, and set in motion the unending struggles for water rights that continue to play out today against the backdrop of devastating drought.[52]

Watkins's photographs did not make much difference in the case's proceedings, it would seem; it was unclear exactly what his images were meant to "prove."[53] The very instability of Watkins's images in their intended role suggests how the broader struggle in *Lux v. Haggin* was less over water rights and more over definitions: what is a river, and how can such questions be described or argued?[54] The entire enterprise of agriculture in this region depended on it. In planning for irrigation in the Great Valley of California, state officials were very precise about the amount of water that the Kern River would provide, converted into measurement in so many inches of water spread over so many acres of land—something that could be quantified and made regular.[55]

Watkins would return to the region in 1886, this time with Haggin as his client and with the aim, as Tyler Green puts it, to show that there was "*so much water*."[56] In his photographs Watkins marshaled all of the clarity and excessively sensitive registration of his enormous glass-plate negatives to demonstrate the way Haggin's irrigation canals cut blithely across the otherwise seemingly unmarked expanses of the Kern Valley. His *Calloway Canal, at Poso* seems to dramatize the awkward and unresolved meeting of two languages of landscape: the meandering shape of the bank and clustering of trees characteristic of the picturesque (as we see established by Constable and Hakewill in their work) and the resolutely nonorganic line of the canal as it stretches toward the horizon (fig. 2.17). One is a language of the local, varied, and limited; the other a language of the displaced, unchanging, and infinite. This is to say, it is a language of the grid, which was the organizing principle ordering the expropriation of land in the Kern Valley and its subsequent remaking under the sign of industrial agriculture.[57] Such a photograph strains to record a landscape in which a cartographic delineation of landscape from above, with its language of objectivity and fixity, has been grafted upon real space (fig. 2.18). Yet Watkins's massive glass-plate negatives cannot trace the canal as it recedes into the distance; some human scale of perception has been disturbed, analogous to the way in which the still surface of the water is turned, in the duration of the image's exposure, almost to a sheet of mirrored glass.

The alteration of water's states—between a mutable, moving substance and a stilled, even abstracted image—is staged in another of Watkins's images of the Calloway Canal and its hydraulic infrastructure that managed the delivery of the Kern River's water to distant alfalfa fields (fig. 2.19). In this bifurcated image, the repetitive gridded structure of the wooden weir, which manages the level of water and diverts it into the canal, comes to define a passage between formlessness and form. We see the water before it passes through the weir, stilled and abstracted; on the other side, having been rinsed through the weir, the water has become active and agitated, dispersing the energies contained in its previously stilled state.[58] In a context such as the court case, where Watkins was called upon to demonstrate photography's capacities to produce stable knowledge about the environment, the weir also takes on a metaphoric relation to the photographic act. Like the weir, a camera transforms the unruly objects of the world into a

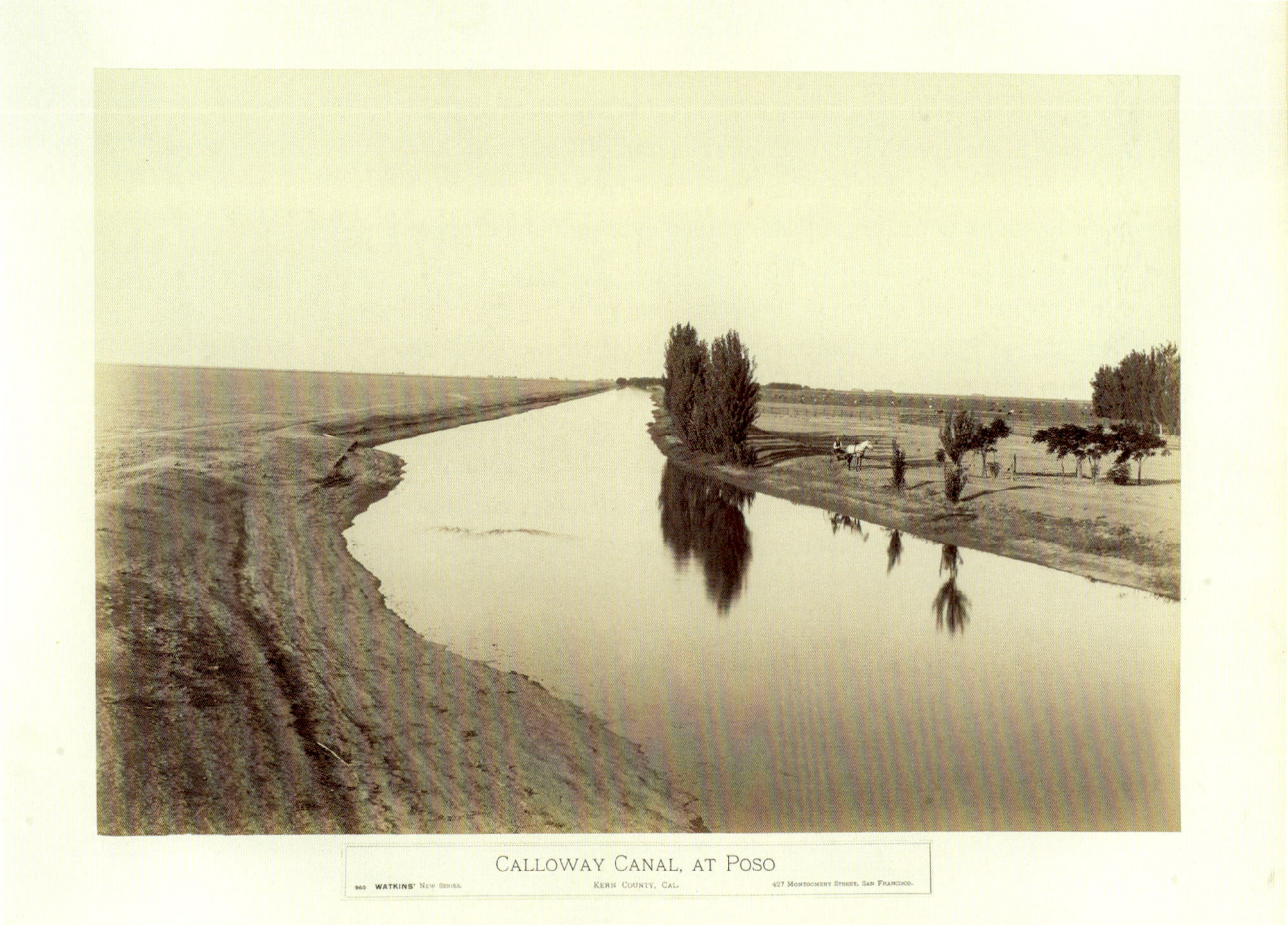

CALLOWAY CANAL, AT POSO

963 WATKINS' NEW SERIES. KERN COUNTY, CAL. 427 MONTGOMERY STREET, SAN FRANCISCO.

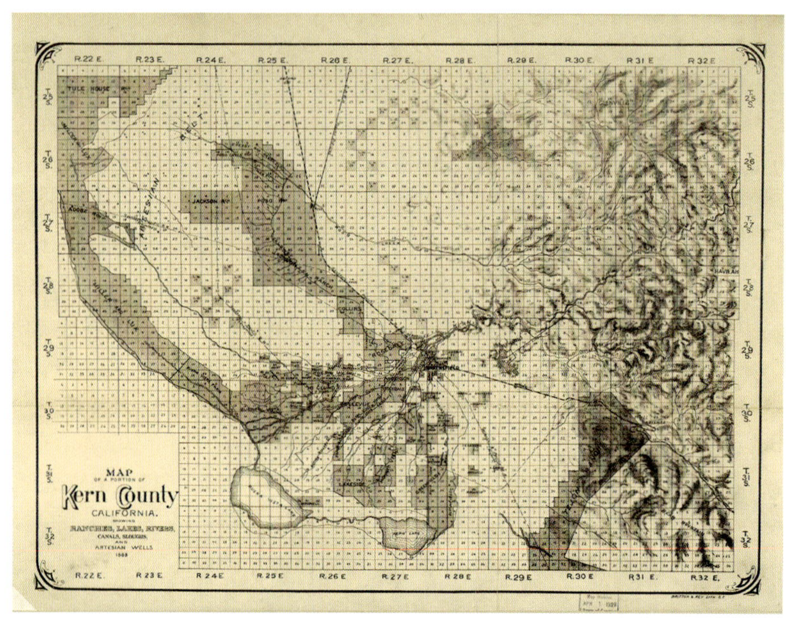

Fig. 2.17. "Calloway Canal, at Poso, Kern County, Cal.," albumen print [19 × 25 in. (48.5 × 64 cm)] in Carleton Watkins's *Photographic Views of Kern County, California* (San Francisco: Carleton Watkins, c. 1881–89), plate 37. The Huntington, 137500.

Fig. 2.18. Britton and Rey, *Map of a Portion of Kern County, California, Showing Ranches, Lakes, Rivers, Canals, Sloughs, and Artesian Wells*, 1888. Lithograph, 14⅛ × 19¼ in. (36 × 49 cm). Library of Congress Geography and Map Division, G4363.K4 1888 .B7.

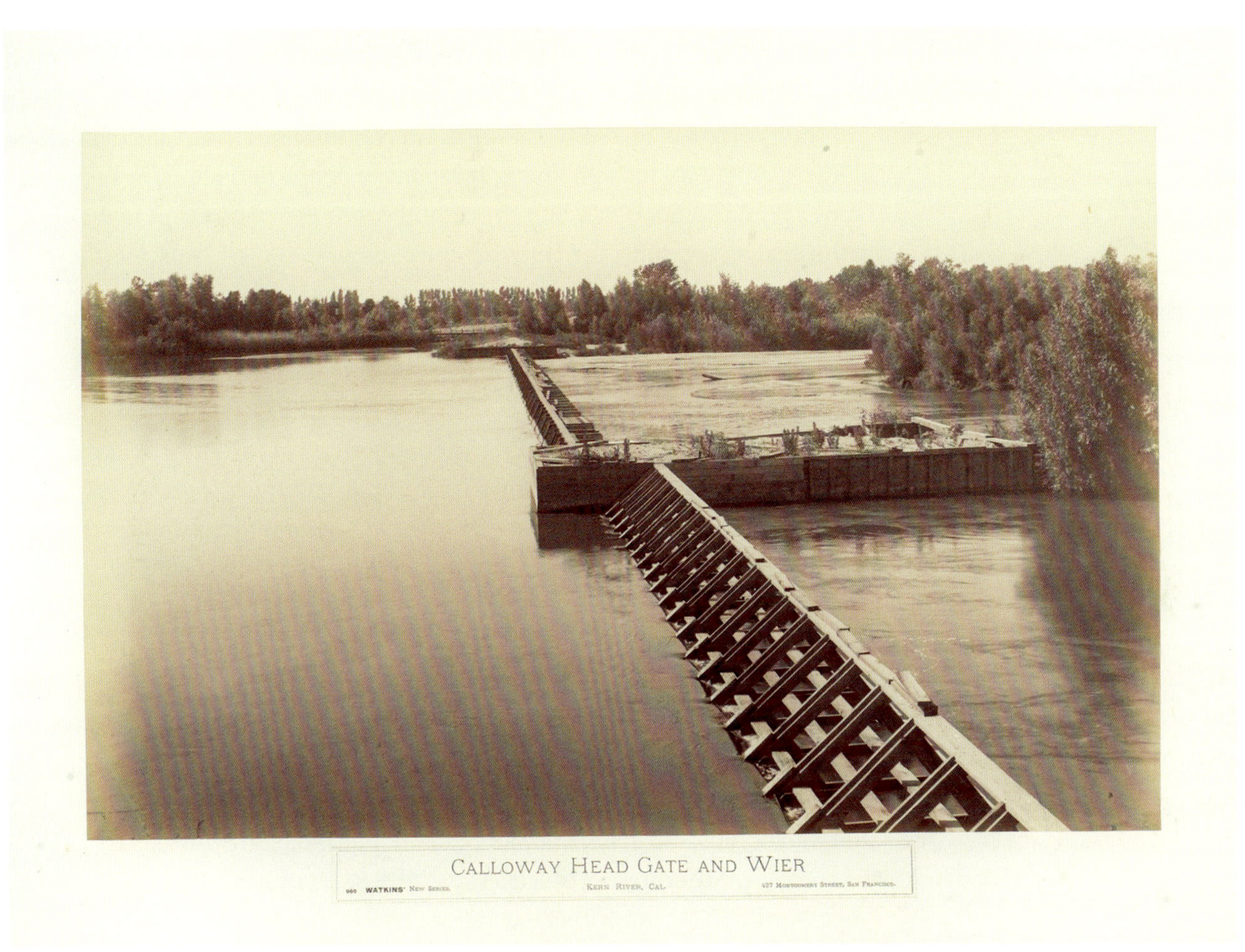

Fig. 2.19. "Calloway Head Gate and Wier, Kern River, Cal.," albumen print [19 × 25 in. (48.5 × 64 cm)] in Carleton Watkins's *Photographic Views of Kern County, California* (San Francisco: Carleton Watkins, c. 1881–89), plate 31. The Huntington, 137500.

distributable and consistent form in a way that might erase the traces by which the image was made.

This is a form of material manipulation very different from Constable's handling of paint. Watkins's photograph depended on the deft handling of photography's liquid components. But that liquid contingency is meant to disappear in the final image, subsumed by what the photographer Jeff Wall calls the "dry," technical, optical elements of the photograph's "hydraulics."[59] We can imagine Watkins, when making these images, carefully applying the liquid collodion that contained the light-sensitive chemicals onto his large glass negatives. An influential manual on the collodion process described the method's difficult negotiations with fluid substance: the operator had to carefully hold the plate while pouring the iodized collodion solution upon it, tilting the glass toward each corner in succession until the glass was covered. One would manipulate the plate back and forth to ensure an even surface, while pouring excess liquid back into the bottle, then immerse the plate in a bath of silver salts to sensitize the plate, and then, as a test of the plate's "sensibility," check to see that the sensitizing liquid "runs off [the glass plate] in an uniform sheet."[60] As a symbolic extension of irrigation and land reclamation, Watkins's prepared glass negatives become a technology of delivery that depends upon the control of liquid. Yet while the recorded surface of the water throws off perceptual data dutifully recorded by the collodion-coated plate, the water does not quite cohere. Instead, as in Watkins's image of the artesian well (see fig. 2.2), it spills over the frame, and almost collapses the photograph's perspective, as if to merge this liquid plane with the surface of the photograph itself.

The water in the Calloway Canal was eventually intended for distribution into smaller irrigation channels ("primary" and "secondary ditches") that would reach the alfalfa fields that grew feed for cattle, sheep, and horses on the area's ranches.[61] As we see in another photograph from Watkins's album, this practice becomes a process of scoring lines in the ground with water seen almost as a medium of drawing (fig. 2.20). The linearity of the diagrams that outlined and prescribed such practices (fig. 2.21) becomes slightly lost in the curving ditches, whose edges are compromised by irregular, sloping banks, hidden from view by masses of grass. The landscape's seemingly elastic scale is belied by the silhouetted figure at right, who tests the surface of the water with a pole, turned away from the cartographic view and immersed in the liquid tactility of the landscape. Here, as in other images, the seemingly agentless power of the irrigation canals is revealed as the product of work. In Kern County, the work of irrigating and harvesting alfalfa was mostly done by Chinese migrant workers in extremely harsh conditions.[62] (A year after Watkins had first photographed this landscape, the 1882 Chinese Exclusion Act was passed to bar the migration of Chinese workers into the United States, part of a broader racist campaign of

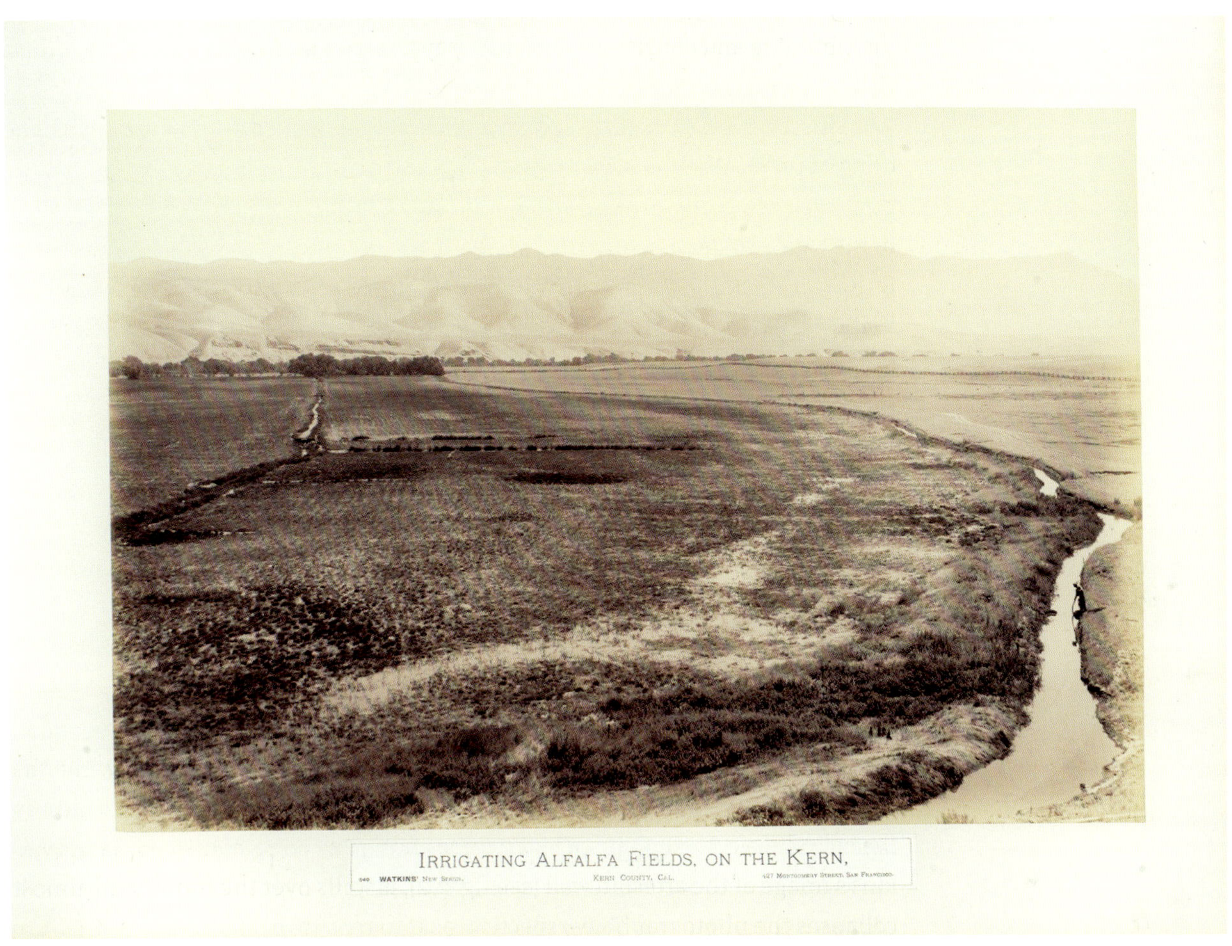

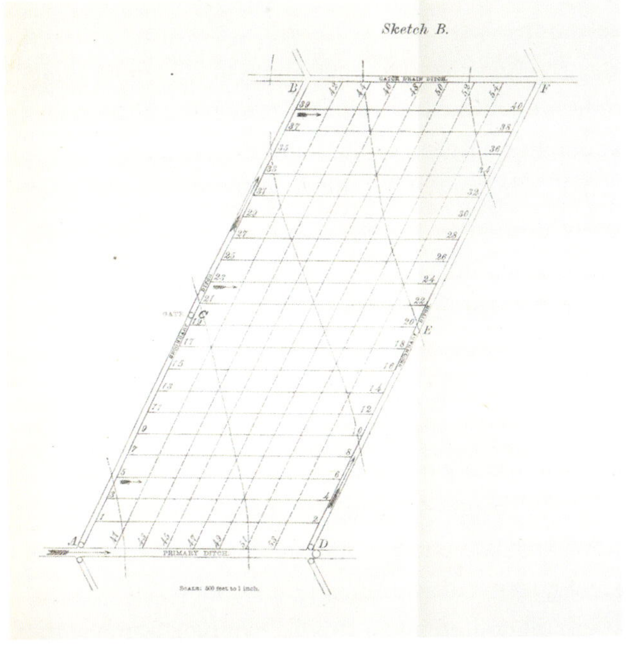

Fig. 2.20. "Irrigating Alfalfa Fields, on the Kern, Kern County, Cal.," albumen print [19 × 25 in. (48.5 × 64 cm)] in Carleton Watkins's *Photographic Views of Kern County, California* (San Francisco: Carleton Watkins, c. 1881–89), plate 24. The Huntington, 137500.

Fig. 2.21. "Sketch B" in *Report of the Board of Commissioners on the Irrigation of the San Joaquin, Tulare, and Sacramento Valleys of the State of California* (Washington, DC: Government Printing Office, 1874), 36. The Huntington, 75071.

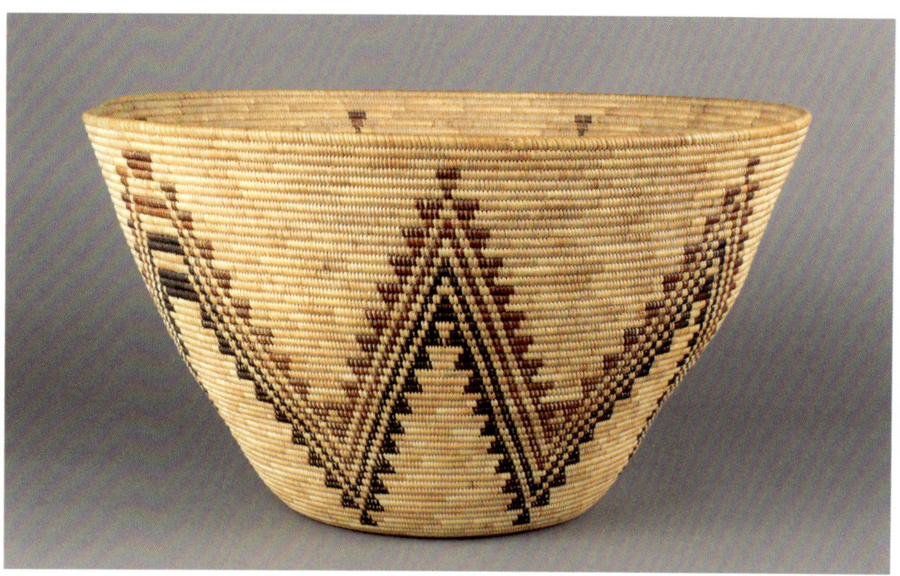

Fig. 2.22. Tule River artist (Tulare), *Basket*, c. 1920. Sedge root, redbud, and blackenfern root filaments, 10¼ × 18¾ in. (26 × 47.6 cm). Courtesy of Portland Art Museum, Portland, Oregon, The Elizabeth Cole Butler Collection, 91.95.28.

violence and restriction.)[63] The disproportion between the scale of the vast irrigation system and the single worker silhouetted against its edge is clear. Yet this figure, whoever he may have been, is also a reminder that, even as Kern County's landscape would be imagined as a site for increasingly mechanized production, it was (as with all of the other spaces discussed in this essay) also a landscape of human life and human power relations. The extraction of natural resources in physical infrastructures was possible because of a social infrastructure for the extraction of human labor. Such systems have persisted in the toxic contamination, drought, and concentration of carceral institutions that continue to limit life in the San Joaquin Valley; the Tule River Tribe continues to fight today for federal recognition of their right to store and use water on its own sovereign lands.[64]

The interconnectivity of natural infrastructure and human life is what the linear, diagrammatic language of resource extraction attempted to displace or obscure. Yet in bringing us up close to the landscape, where the irrigator probes the messy banks of the distribution ditch, Watkins's images show us how such "abstract grids containing . . . raw matter," as the artist Robert Smithson writes in a different context, "are observed as something incomplete, broken, and shattered."[65] Water's strange materiality and its unmanageable relation to fixed, bounded form is part of this; so is the everyday resistance and materiality of work. This is, in part, what we might see in the work of the unidentified Yokuts weaver depicted in Pierce's photograph and in baskets made near Tule River in this period (fig. 2.22). Their making is dependent on a different kind of social and ecological infrastructure: long-term cultivation of plants in an otherwise arid environment, which are then integrated into a supple, complex framework in

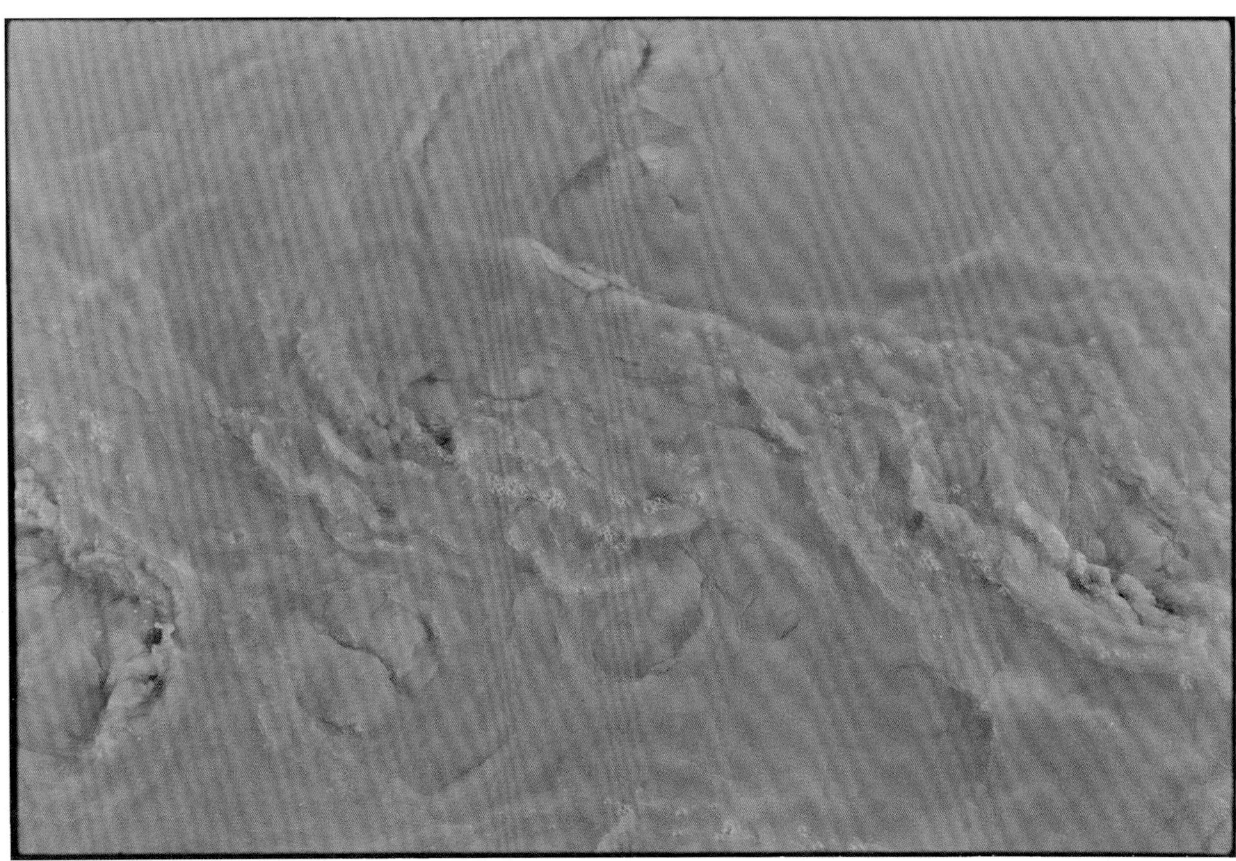

Fig. 2.23. Zoe Leonard, *Al río / To the River* (detail), 2016–2022. Gelatin silver prints, C-prints, and inkjet prints, dimensions variable. Courtesy of the artist, Galerie Gisela Capitain, Cologne, and Hauser and Wirth.

the woven, watertight vessel. Rather than the placeless lines drawn in an irrigation diagram, the complex and nonlinear building up of materials in the process of the basket's weaving points to an ongoing relationship between making and place. Such a relationship to resources is oriented askance from the destructive demands of the industrial and extractive infrastructure that both unfold and dissolve across the tenuous liquid surface of Watkins's photographs.

On the United States–Mexico Border, 2016–22

At the center of Zoe Leonard's photograph we see tender sprays of frothy bubbles, contingent structures floating on shallow crests of gray-brown water (fig. 2.23). The bubbles form amid a liquid topography of swells, waves, shallows, circular eddies, and crevices that cover the surface of the color photograph. This is one of several similar images of the surface of the Rio Grande that form part of Leonard's *Prologue* (2017/2022), which itself is part of a multipart work, *Al río / To the River*, made between 2016 and 2022.[66] These images are eminently photographic in that they show the world to be never the same between any given moment, however quickly it might be perceived or recorded. This is to say, in this water we find a world that is constantly becoming different to itself. *Prologue* is set apart in its medium (color, rather than black-and-white) from Leonard's *Al río*, which moves in a loping, recursive fashion along the Rio Grande and (therefore) the border between Mexico and the United States. The river is one character in these images; another is the "wall" along this border that the U.S. government was building in the years Leonard made these photographs, under a cloud of vituperative and racist rhetoric. Against the strange, attenuated metal slats that compose this wall—forming apertures through which the landscape emerges, disjointed—the river takes its own course through the world, becoming the backdrop and enveloping surround for everyday life: swimming, walking, commuting, and, for the birds, taking flight. The photographs have a buried, quiet narrative: the repeated indifference of the material world to borders, through and over which ecologies of life (human and otherwise) move, despite the everyday violence that attempts to uphold them.

These close-up images of the river itself stand at an elliptical remove from Leonard's other images of the landscape along the Rio Grande. They are a close study of the river's material poetics, patently illegible yet absorbing in their unintended grace. Like the other waterways discussed here, the Rio Grande has been made to do the work of infrastructure—in its case, as a physical and symbolic border. Leonard's photograph allows us to look at the river otherwise and to imagine how communities and attachments are formed outside or alongside political or military jurisdiction.[67] In Leonard's photograph we find no trace of the linear language of cartography, through which it was imagined that natural forms would follow the script set for them under social and political regimes of power. Yet it is not quite clear that the swirling surface of the water records a kind of "resistance" to those regimes either. To ask this river, or ecological forms at large, to do the work of resisting the instrumental and destructive projects of human life would perhaps be asking too much of nature, yet again.

Nicholas Robbins is Lecturer in British Art, 1700–1900, at University College London. He writes about intersecting histories of art, science, and environment in the modern Atlantic world.

Notes

1. This is one of a large number of views Watkins made for land developers in Kern County; the appended printed caption from which I quote likely was supplied by his clients. Originally bound into an album, these images were disbanded and remounted by the Library of Congress after they were acquired in 1929. See the description of this collection: "Photographic Views of Kern County, California," https://www.loc.gov/item/2005690316/.
2. Bruce Bernstein, "Western Baskets," in David W. Penney, *Indigenous Beauty: Masterworks of American Indian Art from the Diker Collection* (New York: Skira Rizzoli in association with American Federation of Arts, 2015), 80. On the Tule River Tribe and Reservation, see Gelya Frank and Carole Goldberg, *Defying the Odds: The Tule River Tribe's Struggle for Sovereignty in Three Centuries* (New Haven: Yale University Press, 2010). On the practices of ecological management that underpin basketry in this region, including the cultivation of sedges along waterways, see M. Kat Anderson, *Tending the Wild: Native American Knowledge and the Management of California's Natural Resources* (Berkeley: University of California Press, 2005), 187–208. For more information about ongoing practices of basket making in Yokuts and Mono communities, including interviews with basketmakers, see Lorrie Planas, "Valley and Foothills: Yokuts and Mono Basketmakers," in Linda E. Dick et al., *Strands of Time: Yokuts, Mono and Miwok Basketmakers* (Fresno, CA: Fresno Metropolitan Museum, 1988), 10–16. This discussion of the relationship between settler-colonial landscape imagery and Indigenous use of natural resources in basketry and other objects is indebted to Julia Lum, "Fire-Stick Picturesque: Landscape Art and Early Colonial Tasmania," *British Art Studies* 10 (November 2018), https://britishartstudies.ac.uk/issues/issue-index/issue-10/fire-stick-picturesque.
3. In focusing on fresh water, this essay follows the model pursued in Sarah Thomas and Natasha Eaton, "Swollen Detail, or What a Vessel Might Give: Agostino Brunias and the Visual and Material Culture of Colonial Dominica," *Atlantic Studies* 19, no. 1 (2022): 60–85.
4. Lauren Berlant, "The Commons: Infrastructures for Troubling Times," *Environment and Planning D: Society and Space* 34, no. 3 (2016): 393.
5. Brian Larkin, "The Politics and Poetics of Infrastructure," *Annual Review of Anthropology* 42 (2013): 335–36.
6. Since both art and infrastructure rely on the abstraction of "nature" as something separate from human life and social organization, as Raymond Williams reminded us fifty years ago, "there is more similarity than we usually recognize between the industrial entrepreneur and the landscape gardener." Raymond Williams, "Ideas of Nature" (1972), in *Problems in Materialism and Culture* (London: Verso, 1980), esp. 75–85 (quote on 81).
7. See Stephen Daniels, "Liquid Landscape: Southam, Constable, and the Art of the Pond," *British Art Studies* 10 (2018), https://www.britishartstudies.ac.uk/issues/issue-index/issue-10/liquid-landscape.
8. This emerges in a comparison between Constable and J. M. W. Turner, whose work is described as "all heat": *Literary Gazette* (May 14, 1831): 315, quoted in Judy Crosby Ivy, *Constable and the Critics, 1803–1837* (Woodbridge, UK: Boydell Press in association with the Suffolk Records Society, 1991), 150–51.
9. *The Champion* (March 7, 1819): 156, quoted in Ivy, *Constable and the Critics*, 80.
10. See John Phillips, *The General History of Inland Navigation*, 4th ed. (London: C. and R. Baldwin, 1803); [Joseph Priestley], *Historical Account of the Navigable Rivers, Canals, and Railways, of Great Britain, as a reference to Nichols, Priestley and Walker's New Map of Inland Navigation, Derived from Original and Parliamentary Documents in the Possession of Joseph Priestly, Esq.* (London: Longman, Rees, Orme, Brown & Green, 1831); Charles Hadfield, *British Canals*, 4th ed. (Newton Abbot, UK: David and Charles, 1969).
11. See Alastair Smart and Attfield Brooks, *Constable and His Country* (London: Elek, 1976); Michael Rosenthal, *Constable: The Painter and His Landscape* (London: Yale University Press, 1983).
12. For an overview, see J. S. Hull, "The River Stour Navigation Company," *Proceedings of the Suffolk Institute of Archaeology* 32, no. 3 (1972): 221–54, including a discussion of the goods transported (see p. 226).
13. Susanna Cole, "Space into Time: English Canals and English Landscape Painting, 1760–1835" (PhD diss., Columbia University, 2013).
14. On the Bridgewater Canal and the so-called Barton Flyover crossing the Irwell, see Cole, "Space into Time," 36–49, esp. 45–49.

15. See Priestly, *Historical Account,* 632–33.
16. See Hull, "River Stour Navigation Company," 221–54.
17. Ann Bermingham, "Reading Constable," *Art History* 10, no. 1 (March 1987): 38–58.
18. Ann Bermingham, *Landscape and Ideology: The English Rustic Tradition, 1740–1860* (Berkeley: University of California Press, 1986), 87–156.
19. See Bermingham, *Landscape and Ideology,* 136–47; *Constable: The Great Landscapes,* ed. Anne Lyles (London: Tate Publishing, 2006). Other key motifs, like the water meadows in Salisbury, were similarly sites of intensive management. See Charles Watkins, "Landscape Management," in *In Focus: Salisbury Cathedral from the Meadows Exhibited 1831 by John Constable,* ed. Amy Concannon (London: Tate Research Publication, 2017), https://www.tate.org.uk/research/in-focus/salisbury-cathedral-constable/landscape-management, accessed June 2, 2023.
20. Constable to John Fisher, October 23, 1821, in *John Constable's Correspondence,* ed. R. B. Beckett, 6 vols. (Ipswich, UK: Suffolk Records Society, 1968), 4:77.
21. See John Barrell, *The Dark Side of the Landscape: The Rural Poor in English Painting, 1730–1840* (Cambridge: Cambridge University Press, 1983), 131–64. As Bermingham writes, the "canal, a sign of his family's involvement in the industrial transformation of the countryside, became a sign as well of Constable's own commercial exploitation of the countryside in the six-footers. The fact that the canal looked like part of the *natural* configuration of the area both occulted the dramatic transformation of the countryside by capital and justified Constable's own representation of it as 'rustic landscape.' " Bermingham, *Landscape and Ideology,* 138.
22. See Humphrey W. Woolrych, *A Treatise of the Law of Waters,* 2nd ed. (London: William Benning, 1851), 1–3, 40–49; Louis Houck, *A Treatise on the Law of Navigable Rivers* (Boston: Little, Brown, 1868); and Hadfield, *British Canals,* 58–62.
23. The "A. Constable, Esq." on the map refers to Abram Constable, the artist's brother, who took over the family's businesses following Golding Constable's death in 1816.
24. For one historian, "the invention of the pound lock was almost as important for economic development as the discovery of steam power." T. S. Willan, *River Navigation in England, 1600–1750* (1936; London: Frank Cass, 1964), 94.
25. This letter from Abram Constable to the commissioners of the River Stour Navigation Company is quoted in Ian Fleming-Williams, *John Constable: A Master Draughtsman* (London: Dulwich Picture Gallery, 1994), 206. Constant repair was required to the Stour's locks and banks. John Constable would, on his father's behalf, attend a River Stour Navigation Company meeting in 1808 to discuss the "ruinous" condition of towpaths and the silting up of the river below Flatford Lock, and to report that the bank around Dedham Lock was "so thin and weak . . . as to be dangerous, for Navigators." River Stour Navigation Company commissioners' minute book, cited in Ian Fleming-Williams, *Constable and His Drawings* (London: Philip Wilson, 1990), 112. On the various repairs and improvements to the Navigation, undertaken in fits and starts, see Hull, "River Stour Navigation Company," 221–54.
26. On the history of improvement and its aesthetic underpinnings, see Vittoria Di Palma, *Wasteland: A History* (New Haven: Yale University Press, 2014).
27. See Graham Reynolds, *The Later Paintings and Drawings of John Constable,* 2 vols. (London: Paul Mellon Centre for Studies in British Art, 1984), 1:30n7.
28. Da Vinci's work on canal architecture and hydraulics is discussed in John Sidney Hawkins, "A New Life of the Author," in *A Treatise on Painting by Leonardo da Vinci,* trans. John Francis Rigaud (London: J. Taylor, 1802), xiii–lxxiv. Constable may also have known about da Vinci's recently transcribed and translated scientific manuscripts, including descriptions of the action of water and the ways that rivers shaped landforms: see, e.g., J. B. Venturi, "Extracts from the Manuscripts of Leonardo da Vinci," *A Journal of Natural Philosophy, Chemistry, and the Arts* 2 (May 1798): 84–93. On Leonardo's drawings of water, see Ernst H. Gombrich, "The Form of Movement in Water and Air," in *Leonardo's Legacy: An International Symposium,* ed. C. D. O'Malley (Berkeley: University of California Press, 1969), 171–204.
29. See Andreas Malm, *Fossil Capital: The Rise of Steam Power and the Roots of Global Warming* (London: Verso, 2016), esp. 37–57, 77–95.
30. Sidney W. Mintz, *Sweetness and Power: The Place of Sugar in Modern History* (1985; New York: Penguin, 1986), 47.
31. Ideologies of climate partitioned the world into torrid, temperate, and frigid "zones" that would, in time, answer to the aims of empire. In an art historical context, see Kay Dian Kriz, *Slavery, Sugar, and the Culture of Refinement: Picturing the British West Indies, 1700–1840* (London: Yale

University Press for the Paul Mellon Centre for Studies in British Art, 2013), 157–93; Nicholas Robbins, "John Constable, Luke Howard, and the Aesthetics of Climate," *Art Bulletin* 103, no. 2 (2021): 50–76. For an overview of Caribbean environmental history, see Philip D. Morgan, "The Caribbean Environment to 1850," in Philip D. Morgan, John R. McNeill, Matthew Mulcahy, and Stuart B. Schwartz, *Sea and Land: An Environmental History of the Caribbean* (Oxford: Oxford University Press, 2022), 19–129.

32. Key accounts include Geoff Quilley, "Pastoral Plantations: The Slave Trade and the Representation of British Colonial Landscape in the Late Eighteenth Century," in *An Economy of Colour: Visual Culture and the Atlantic World, 1660–1830*, ed. Geoff Quilley and Kay Dian Kriz (Manchester: Manchester University Press, 2003), 106–28; Jill Casid, *Sowing Empire: Landscape and Colonization* (Minneapolis: University of Minnesota Press, 2005); *Art and Emancipation in Jamaica: Isaac Mendes Belisario and His Worlds*, ed. Tim Barringer, Gillian Forrester, and Bárbaro Martínez-Ruiz (New Haven: Yale University Press, 2007); Kriz, *Slavery, Sugar, and the Culture of Refinement*; Charmaine A. Nelson, *Slavery, Geography, and Empire in Nineteenth-Century Marine Landscapes of Montreal and Jamaica* (London: Routledge, 2017); and Tim Barringer, "Land, Labor, Landscape: Views of the Plantation in Victorian Jamaica," in *Victorian Jamaica*, ed. Barringer and Wayne Modest (Durham, NC: Duke University Press, 2018), 282–321.
33. The description of the plantation's former owner, Zachary Bayly, could suffice to limn this ideology: "To whom the endowments of Nature rendered those / Of Art superfluous." Quoted in James Hakewill, letterpress to *A Picturesque Tour of the Island of Jamaica, from Drawings Made in the Years 1820 and 1821* (London: Hurst and Robinson, 1824), n.p.
34. See Hakewill, letterpress to *A Picturesque Tour of the Island of Jamaica*, n.p. In the last affordance of Trinity Estate he mentions, Hakewill describes the usually unseen shadow of the plantation's monocultural cash-crop production of sugar and other goods for trade: the provision grounds in which enslaved workers grew their own sustenance and raised crops for market. Such spaces are often cited as sites of resistance to the violent theft of life and labor that would otherwise define this world.
35. William Beckford, *A Descriptive Account of the Island of Jamaica* (London: T. and J. Egerton, 1790), 1:358.
36. This is how Beckford describes the communal gathering of enslaved workers at Christmas; see Beckford, *Descriptive Account*, 1:391. On the representation of labor in the eighteenth- and nineteenth-century Caribbean, see Mia L. Bagneris, *Colouring the Caribbean: Race and the Art of Agostino Brunias* (Manchester: Manchester University Press, 2018), 92–135.
37. See B. W. Higman, *Plantation Jamaica, 1750–1850: Capital and Control in a Colonial World* (Mona: University of the West Indies Press, 2005), 180–90; Veront M. Satchell, "Early Use of Steam Power in the Jamaican Sugar Industry, 1768–1810," *Transactions of the Newcomen Society* 67, no. 1 (1995): 221–31; Aaron Graham, "Technology, Slavery, and the Falmouth Water Company of Jamaica, 1799–1805," *Slavery and Abolition* 39, no. 2 (2018): 315–32; Hayden F. Bassett, "Plantation Roads and the Impositions of Infrastructure: An Archaeology of Movement at Good Hope Estate, Jamaica," *Journal of African Diaspora Archaeology and Heritage* 11, no. 1 (2022): 48–73; Ramesh Mallipeddi, "Roads, Bridges, and Ports: Infrastructures of Plantation Agriculture in the British Caribbean, 1627–1840," in *The Aesthetic Life of Infrastructure*, ed. Kelly Mee Rich, Nicole M. Rizzuto, and Susan Zieger (Evanston, IL: Northwestern University Press, 2023), 23–42.
38. See Bryan Edwards, *The History, Civil and Commercial, of the British Colonies in the West Indies*, 3rd ed. (London: John Stockdale, 1801), 2:261; Beckford, *Descriptive Account*, 1:151.
39. Hakewill, letterpress to *A Picturesque Tour of the Island of Jamaica*, n.p. For a discussion and comparison with surviving maps of the estate, see B. W. Higman, *Jamaica Surveyed: Plantation Maps and Plans of the Eighteenth and Nineteenth Centuries* (Mona: University of the West Indies Press, 1988), 116–18. On "improvement" and the Jamaican plantation landscape, including a discussion of Trinity Estate, see Louis P. Nelson, *Architecture and Empire in Jamaica* (New Haven: Yale University Press, 2016), 97–130, esp. 106–9.
40. Édouard Glissant, *Poetics of Relation*, trans. Betsy Wing (1990; Ann Arbor: University of Michigan Press, 1997), 64, 67.
41. See Richard H. Grove, *Green Imperialism: Colonial Expansion, Tropical Island Edens, and the Origins of Environmentalism, 1600–1860* (Cambridge: Cambridge University Press, 1995). See also Anna Arabindan-Kesson, "Transmission and Transfer: Plantation Imagery and Medical Management in the British Empire," *Art History* 45, no. 3 (June 2022): 472–97, esp. 482–90.
42. Edward Long, *The History of Jamaica; or, General Survey of the Antient and Modern State of That Island* (London: T. Lowndes, 1774), 1:465–66. See also Mallipeddi, "Roads, Bridges, and Ports,"

35–36. This draws on a language that saw the circulation of people and goods in a city as akin to the circulation of blood in the body; see Richard Sennett, *Flesh and Stone: The Body and the City in Western Civilization* (New York: W. W. Norton, 1994), 255–70.

43. Larkin, "Politics and Poetics of Infrastructure," 329.
44. J. Stewart, *A View of the Past and Present State of the Island of Jamaica, with Remarks on the Moral and Physical Condition of the Slaves, and on the Abolition of Slavery in the Colonies* (Edinburgh: Oliver and Boyd, 1823), 33.
45. Edwards, *History*, 2:75–79. See Vincent Brown, *Tacky's Revolt: The Story of an Atlantic Slave War* (Cambridge, MA: Harvard University Press, 2022).
46. Vincent Brown, *The Reaper's Garden: Death and Power in the World of Atlantic Slavery* (Cambridge, MA: Harvard University Press, 2008), 8.
47. The question of how water shaped, and was shaped by, everyday life on plantations is outside of this account, but is explored in Thomas and Eaton, "Swollen Detail," 60–85.
48. Thomas Clarkson, *The History of the Rise, Progress, and Accomplishment of the Abolition of the African Slave-Trade by the British Parliament* (London: Longman, Hurst, Rees, and Orme, 1808), 1:259, 1:32.
49. Here I follow Marcus Wood, *Blind Memory: Visual Representations of Slavery in England and America, 1780–1865* (Manchester: Manchester University Press, 2000), 1–6.
50. This period of Watkins's career has been recently treated with acute attention by Tyler Green: see *Carleton Watkins: Making the West American* (Berkeley: University of California Press, 2018), 395–412, 421–50. See also Richard Steven Street, *A Kern County Diary: The Forgotten Photographs of Carleton E. Watkins, 1881–1889* (Bakersfield, CA: Kern County Museum, 1983); Peter E. Palmquist, *Carleton E. Watkins: Photographer of the American West* (Albuquerque: University of New Mexico Press, 1983), 75–78; Weston Naef and Christine Hult-Lewis, *Carleton Watkins: The Complete Mammoth Photographs* (Los Angeles: J. Paul Getty Museum, 2011), 411–33; and Christine Hult-Lewis, "The Mining Photographs of Carleton Watkins, 1858–1891, and the Origins of Corporate Photography" (PhD diss., Boston University, 2011), 231–38.
51. In one then-recent, violent episode in the 1860s, during wars between Native and U.S. combatants, the U.S. Army Captain Moses A. McLaughlin had murdered thirty-five men from Tübatulabal and Owens Valley Paiute communities at Keysville on the Kern River, upstream from Bakersfield; see Benjamin Madley, *An American Genocide: The United States and the California Indian Catastrophe, 1846–1873* (New Haven: Yale University Press, 2016), 314.
52. See Donald J. Pisani, *From the Family Farm to Agribusiness: The Irrigation Crusade in California and the West, 1850–1931* (Berkeley: University of California Press, 1984), esp. 191–249; Donald Worster, *Rivers of Empire: Water, Aridity, and the Growth of the American West* (New York: Pantheon Books, 1985), esp. 103–11; Eric. T. Freyfogle, "*Lux v. Haggin* and the Common Law Burdens of Modern Water Law," *University of Colorado Law Review* 57 (1986): 485–525.
53. In the testimony he gave in court, Watkins was pressed about the nature of the evidence his image might be able to provide. He was asked, for example, if one could measure the width of a canal or correctly trace the edges of a stream's banks using a photograph. While Watkins insisted in his testimony that you might, this was not convincing to the court. It was rather to other forms of material evidence—like the surveyor's measuring chain—to which the trial turned to confirm the dimensions and form of Kern County's waterways. See "Testimony of C. E. Watkins," in Lux v. Haggin, 69 Cal. 255, 10 P. 674 (1886): 2:577–81, https://archive.org/details/car_002410; and "Testimony of S. W. Wibble [*sic*]," in *Lux*, 69 Cal. 255, 10 P. 674 (1886), 2:583. On photography's evidentiary status in legal contexts, see Jennifer Mnookin, "The Image of Truth: Photographic Evidence and the Power of Analogy," *Yale Journal of Law and the Humanities* 10, no. 1 (1998): 1–74.
54. See David Igler, *Industrial Cowboys: Miller and Lux and the Transformation of the Far West, 1850–1920* (Berkeley: University of California Press, 2001), esp. 103–11.
55. "From rough observations of the actual discharge of Kern River near the end of May, 1873, it was found to be equal to a depth of 1 1/2 inches per month from the whole area of cachement of 2,400 square miles. This could give a depth of 3 inches for irrigation over 1,200 square miles, or 768,000 acres, which is larger than its natural irrigation district." Barton Stone Alexander, George Davidson, and George Henry Mendell, *Report of the Board of Commissioners on the Irrigation of the San Joaquin, Tulare, and Sacramento Valleys of the State of California* (Washington, DC: Government Printing Office, 1874), 24.
56. Green, *Carleton Watkins*, 424 (emphasis in the original). Not all of the images made in Kern County in the late 1880s use the large wet collodion plates; some, as fig. 2.2, were made with a

new "dry plate" technology in which glass negatives were sensitized in advance and could be more easily transported; see Green, *Carleton Watkins*, 422–23.

57. On this image, and on the grid in Watkins's Kern images, see Green, *Carleton Watkins*, 425–27, 449–50. On the grid and landscape photography in the U.S. West in a different context, see Thomas Cornelius, "Landscape as Grid in Stephen Shore's 'American Surfaces,'" *Burlington Magazine* 165, no. 1442 (May 2023): 522–29.
58. See also the account of this image in Hult-Lewis, "Mining Photographs," 237.
59. Wall has suggested that a "liquid intelligence"—a kind of unconscious formlessness—haunts the otherwise "dry" medium of photography and its "hydraulics" of control. In photography, Wall writes, water must be "controlled exactly and cannot be permitted to spill over the spaces and moments mapped out for it in the process." The "dry" elements of photography—its "optics and mechanics"—must be protected from the "immersion in the incalculable" that water represents in the photographic medium. Jeff Wall, "Photography and Liquid Intelligence," in *Jeff Wall: Selected Essays and Interviews*, ed. Peter Galassi (New York: Museum of Modern Art, 2007), 109–10. On the expanded significance of Wall's formulation, see the group of essays on "Liquid Intelligence," edited by Matthew C. Hunter, in *Grey Room* 69 (Fall 2017): 6–136.
60. Frederick Scott Archer, *The Collodion Process on Glass*, 2nd ed. (London: s.n. 1854), 44–46.
61. Pisani, *From the Family Farm to Agribusiness*, 244–45.
62. As Richard Steven Street describes, "the work, which required men to stand knee-deep and barefoot in cold water for many hours under the hot sun, was painfully uncomfortable." Street, *Beasts of the Field: A Narrative History of California Farmworkers, 1769–1913* (Stanford, CA: Stanford University Press, 2004), 324. See also Igler, *Industrial Cowboys*, 122–46.
63. See Erika Lee, *The Making of Asian America: A History* (New York: Simon and Schuster, 2015), 89–108; Beth Lew-Williams, *The Chinese Must Go: Violence, Exclusion, and the Making of the Alien in America* (Cambridge, MA: Harvard University Press, 2018); and the oral histories in William Harland Boyd, *The Chinese of Kern County, 1857–1960* (Bakersfield, CA: Kern County Historical Society, 2002), 33–46.
64. On the San Joaquin Valley and its carceral histories and present, see Ruth Wilson Gilmore and Craig Gilmore, "The Other California," in Ruth Wilson Gilmore, *Abolition Geography: Essays towards Liberation* (London: Verso Books, 2022), 242–58. For more information about the Tule River Tribe's present-day struggle to secure their water rights, see the resources gathered on the tribe's website, https://tulerivertribe-nsn.gov/water/, last accessed August 10, 2023.
65. Robert Smithson, "A Sedimentation of the Mind: Earth Projects" (1968), in *Robert Smithson: The Collected Writings*, ed. Jack Flam (Berkeley: University of California Press, 1996), 110.
66. I saw *Al río / To the River* as it was exhibited at the Musée d'art moderne de Paris (October 15, 2022–January 29, 2023); it comprises three separate parts, *Prologue*, *Al río*, and *Coda*. The work has been published as *Zoe Leonard: Al río / To the River*, ed. Tim Johnson (Berlin: Hatje Cantz, 2022).
67. Here I follow Lauren Berlant's affective conception of infrastructure as "that which binds us to the world in movement and keeps the world practically bound to itself," and therefore as a set of relations. See Berlant, "Commons," 394.

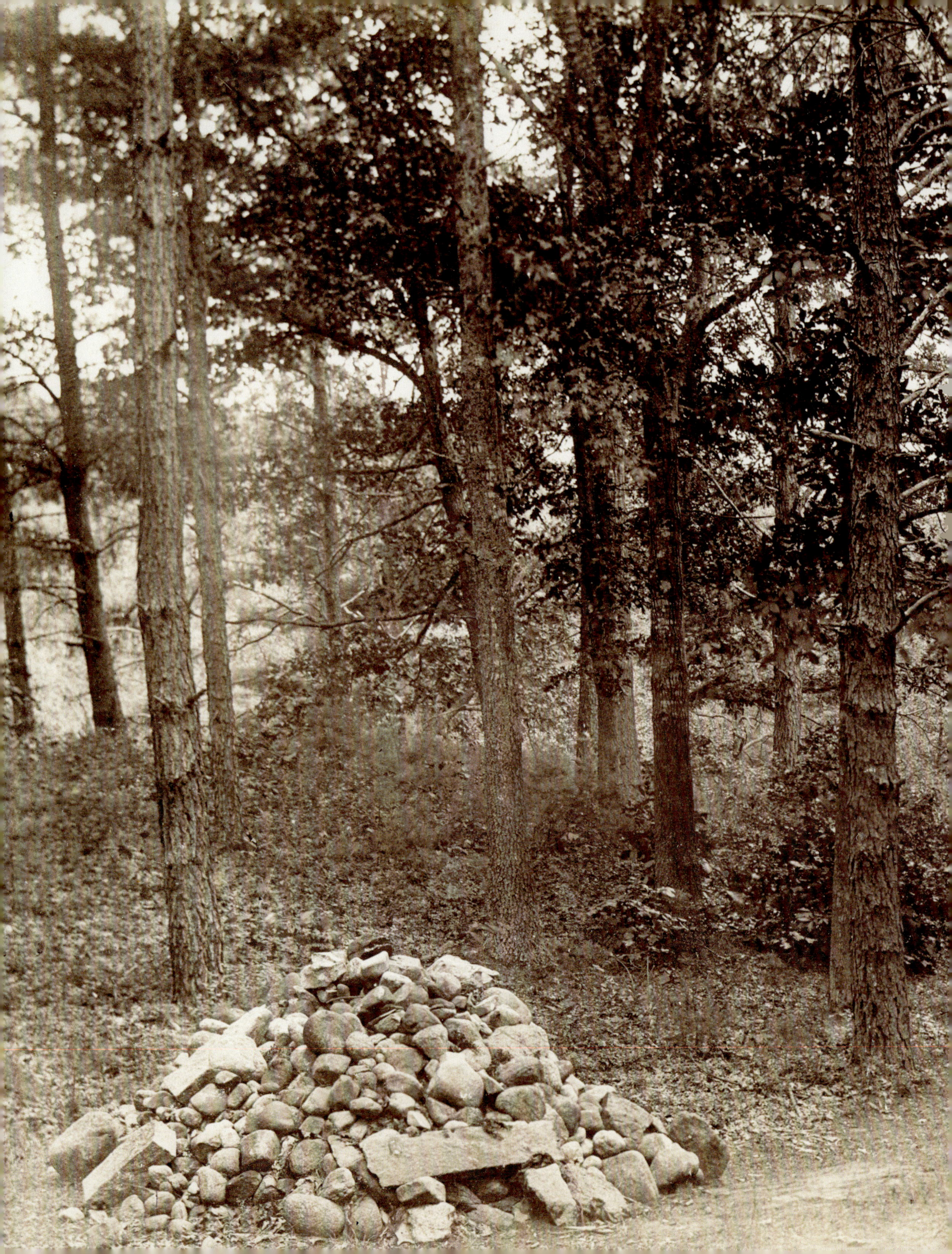

Kristen Case

"They fetch the year about to me"

Thoreau's Ways of Knowing

I perceive that we partially die ourselves through sympathy at the death of each of our friends or near relatives. Each such experience is an assault on our vital force. It becomes a source of wonder that they who have lost many friends still live. After long watching around the sick-bed of a friend, we, too, partially give up the ghost with him, are the less to be identified with the state of things.

. . . When we have experienced many disappointments, such as the loss of friends, the notes of birds cease to affect us as they did.

—Henry David Thoreau, Journal, February 3–5, 1859 (J 11:438–39)

I spend a considerable portion of my time observing the habits of the wild animals, my brute neighbors. By their various movements and migrations they fetch the year about to me. Very significant are the flight of geese and the migration of suckers, etc. But when I consider that the nobler animals have been exterminated here, the cougar, panther, lynx, wolverene, wolf, bear, moose, deer, beaver, turkey, etc., etc., I cannot but feel as if I lived in a tamed and, as it were, emasculated country. Would not the motions of those larger and wilder animals have been more significant still? Is it not a maimed and imperfect nature that I am conversant with? . . . When I think what were the various sounds and notes, the migrations and works, and changes of fur and plumage which ushered in the spring and marked the other seasons of the year, I am reminded that this my life in nature, this particular round of natural phenomena which I call a year, is lamentably incomplete. . . . Many of those animal migrations and other phenomena by which the Indians marked the season are no longer to be observed. I seek acquaintance with nature to know her moods and manners. . . . I take

> infinite pains to know all the phenomena of the spring, for instance, thinking that I have here the entire poem, and then, to my chagrin, I learn that it is but an imperfect copy that I possess and have read, that my ancestors have torn out many of the first leaves and grandest passages, and mutilated it in many places.
>
> —Henry David Thoreau, Journal, March 23, 1856 (J 8:220–21)

On February 3, 1859, Henry David Thoreau devoted an entire page of his journal to a single sentence: "Five minutes before 3PM, Father died." Later that day he makes the following observation: "When we have experienced many disappointments, such as the loss of friends, the notes of birds cease to affect us as they did" (J 11:438–39). It is a revealing connection: personal loss, individual bereavement, entails a more-than-personal kind of pain. For Thoreau, grief changed the season.

At the same time, like many of the writers, artists, and scientists whose work is featured in the *Storm Cloud* exhibition, Thoreau clearly registered the impact of human activity on the natural world. Reflecting on the species loss that already haunted the woods of Concord, Massachusetts, he asked, "Is it not a maimed and imperfect nature that I am conversant with?" The question illustrates the extent of the ecological wound that he and others had begun to recognize. In Thoreau's choice of the word "conversant," however, we also see the intimacy that this woundedness betrays. The grieving listener who registers a change in his responsiveness to birdsong has already developed an acute attunement to it.

Throughout his mature writings, Thoreau sought to be "identified with the state of things" in the natural world, and to "know all the phenomena" of each season. Most often he described this knowing not in terms of expertise, but in terms of acquaintanceship, neighboring, or friendship: "I spend a considerable portion of my time observing the habits of the wild animals, my brute neighbors," he writes. "By their various movements and migrations they fetch the year about to me." In this formulation, it is the nonhuman animals around him that *create* the year for Thoreau, actively carrying its meaning in the visible patterns of their "movements and migrations." This relational sense of and orientation to the more-than-human world meant that, for Thoreau, individual loss and suffering was tied to a sense of alienation from his environment, a feeling of being out of step with nature.[1] Conversely, Thoreau's awareness of environmental damage, species loss, and increasing human alienation from nature during what he called "this restless, nervous, bustling, trivial Nineteenth Century" was for him a source of deeply *personal* bereavement:[2] "This my life in nature, this particular round of natural phenomena which I call a year, is lamentably incomplete."

Corresponding to this interwoven sense of personal and ecological loss was

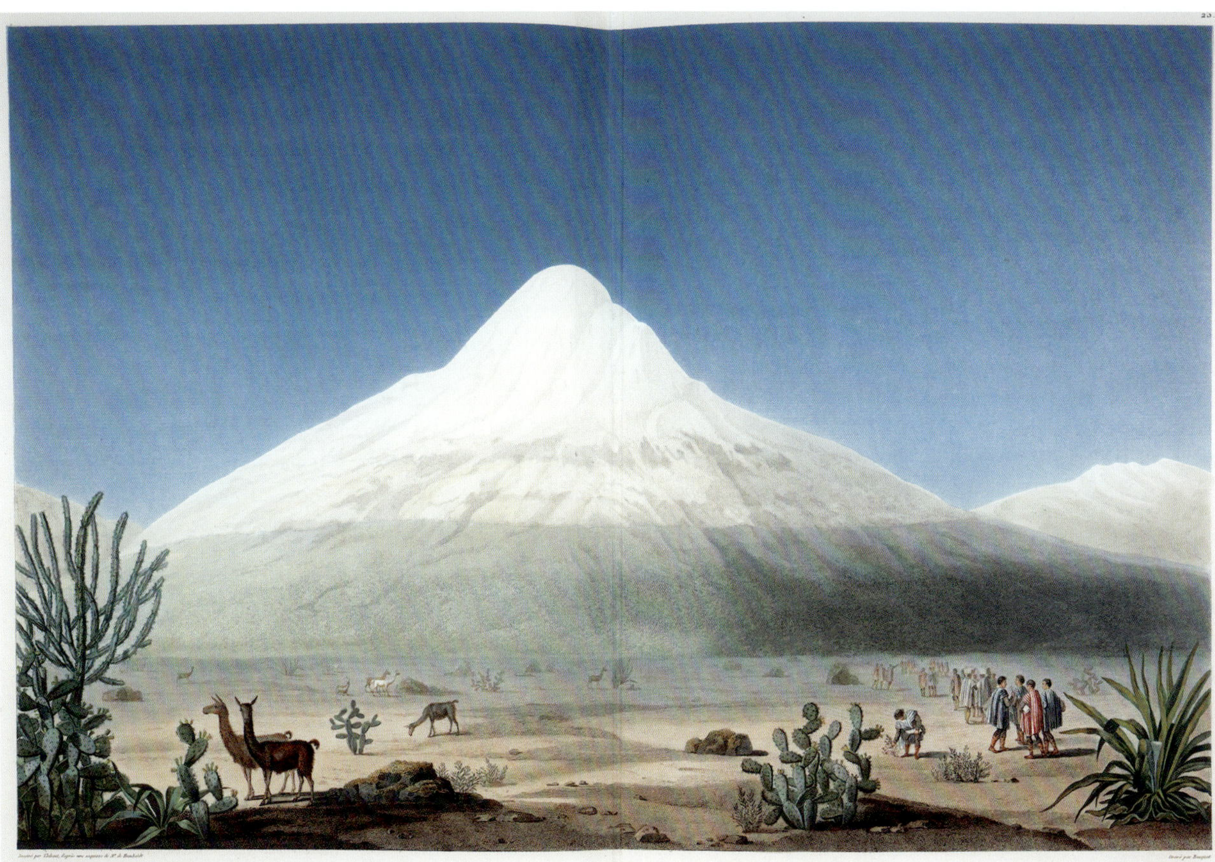

its complement: an understanding of health or "sanity" that resulted from and was reflected by attunement to the more-than-human world. Though the feelings that anchored this orientation seem to have been lifelong, Thoreau made refining his deep sense of this relation the central work of his final years. Influenced not only by naturalists such as Gilbert White but also by the new and forward-looking theories of Alexander von Humboldt and Charles Darwin, Thoreau followed these thinkers in conceiving of the human as an integral part of what Humboldt, in *Views of Nature,* calls "Cosmos: the dance of world and mind, subject and object" (fig. 3.1).[3]

Fig. 3.1. "Chimborazo," frontispiece to Alexander von Humboldt's *Vues des Cordillères, et monumens des peuples indigènes de l'Amérique* (Paris: F. Schoell, c. 1810). The Huntington, 88919.

This essay will trace the emergence of one aspect of Thoreau's late-life practices of attunement to the more-than-human world: the development of his Kalendar, or monthly charts of general phenomena. In the conception and development of these charts, we witness tendencies common to nineteenth-century thought—a growing interest in statistical averages and composite visualizations, a new awareness of deep time, and a reconceptualization of the relation between human and nonhuman elements of the natural world. These charts also reveal the way these interests and ideas intersected with Thoreau's personal experience of living—and dying—in a more-than-human world. This is an essay about a network of practices of being in, observing, and recording that world. The nature of these practices means that it is also an essay about slavery in America, grief, evolution, friendship, and disability.

Part I: Spring 1860

April 19 Surveying JB Moore's farm.

Hear the field sparrow sing on his dry upland, it being a warm day, and see the small blue butterfly hovering over the dry leaves. Toward night, hear a partridge drum. You will hear at first a single beat or two far apart and have time to say, "There is a partridge," so distinct and deliberate is it often, before it becomes a rapid roll.

—Henry David Thoreau, Journal, April 19, 1860 (J 8:252)

In April of 1860, Henry David Thoreau was watching butterflies, listening for sparrows, and once again watching the spring come in. The changing of the season—and especially the arrival of spring—was always a significant occasion for Thoreau. Far from a backdrop to his life, the new season was the main event. In *Walden* he had written,

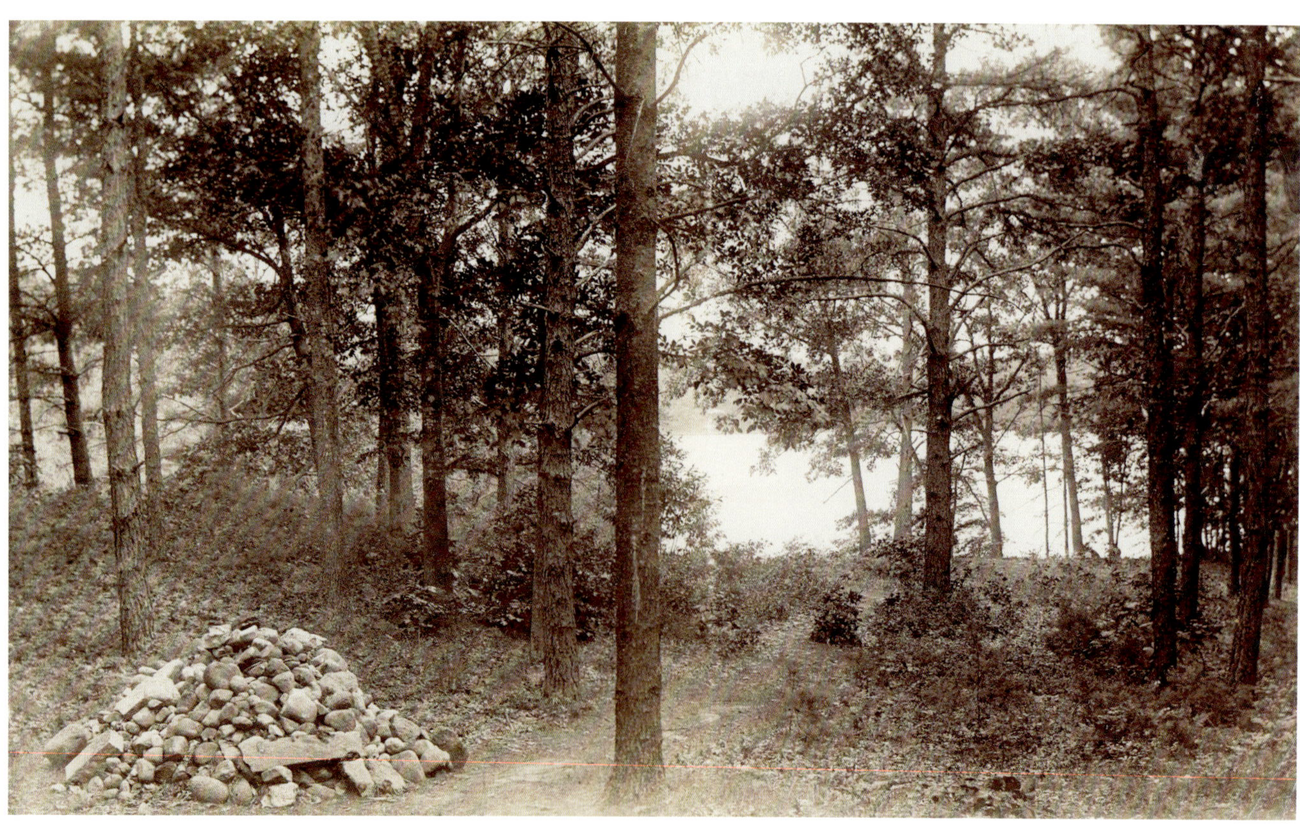

Fig. 3.2. The site of Thoreau's cabin, Walden Pond, ca. 1895. Albumen print, 4½ × 7½ in. (11.4 × 19 cm). The Huntington, Correspondence of Prudence Ward and Anne J. Ward, 1839–1906.

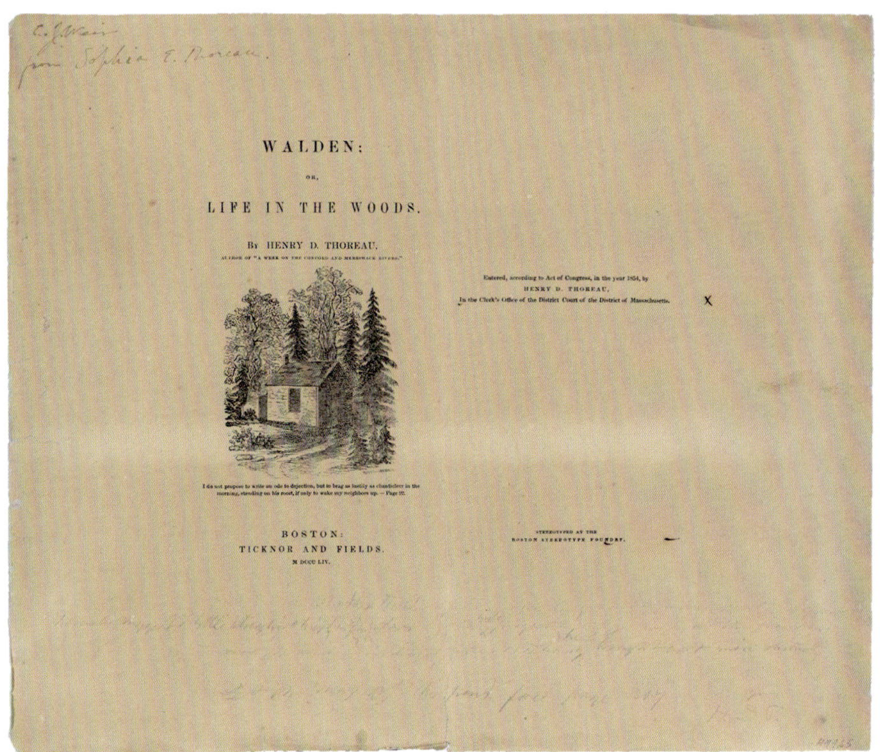

WALDEN;

OR,

LIFE IN THE WOODS.

BY HENRY D. THOREAU,

AUTHOR OF "A WEEK ON THE CONCORD AND MERRIMACK RIVERS."

I do not propose to write an ode to dejection, but to brag as lustily as chanticleer in the morning, standing on his roost, if only to wake my neighbors up. — Page 92.

BOSTON:
TICKNOR AND FIELDS.
M DCCC LIV.

Entered, according to Act of Congress, in the year 1854, by
HENRY D. THOREAU,
In the Clerk's Office of the District Court of the District of Massachusetts.

STEREOTYPED AT THE
BOSTON STEREOTYPE FOUNDRY.

Fig. 3.3. Corrected page proof for the frontispiece to Henry David Thoreau's *Walden; or, Life in the Woods* (Boston: Ticknor and Fields, 1854). The Huntington, mssHM 925.

> One attraction in coming to the woods to live was that I should have leisure and opportunity to see the spring come in. The ice in the pond at length begins to be honey-combed, and I can set my heel in it as I walk. Fogs and rains and warmer suns are gradually melting the snow; the days have grown sensibly longer; and I see how I shall get through the winter without adding to my wood-pile, for large fires are no longer necessary. I am on the alert for the first signs of spring, to hear the chance note of some arriving bird, or the striped squirrel's chirp, for his stores must be now nearly exhausted, or see the woodchuck venture out of his winter quarters (figs. 3.2, 3.3).[4]

Now, years after his experiment at Walden Pond, Thoreau was no less alert for signs of the turning year. Watching for, observing, and recording signs of seasonal change had become a way of being in the world, a daily practice. Charting the days of his own life along the axis of natural time, he became interested in what he called "annual phenomena," signs of the season to which he felt a special connection, and that linked his days to the days of previous years, helping him orient himself in the world, and in time.

In October of 1859 he had written, "For thirty years I have annually observed, about this time or earlier, the freshly erected winter lodges of the musquash along the riverside. . . . This may not be an annual phenomenon to you. It may not be in the Greenwich almanac or ephemeris, but it has an important place in my Kalendar" (J 12:339). The Royal Observatory in Greenwich, England, began

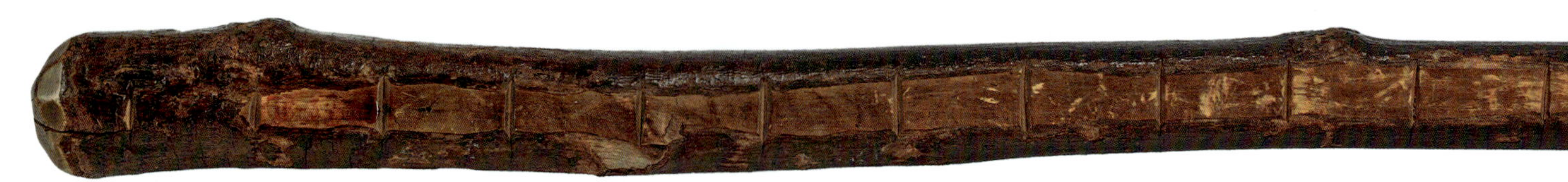

Fig. 3.4. Henry David Thoreau, walking stick (and detail), c. 1853, Concord, Massachusetts. Birch, 1 × 1 × 38 in. (2.5 × 2.5 × 96.4 cm). Courtesy of the Concord Museum, Gift of Lee, Olive, and Earnest Russell, Th34.

publishing its various almanacs in the eighteenth century; however, ephemera, or charts tracking the trajectory of astronomical objects over time, had existed since ancient Mesopotamian civilizations. Thoreau would have been familiar with the ancient Roman ephemera of Ptolemy. Watching for musquash lodges as astronomers watched the movements of the planets in the night sky, Thoreau pondered the possibilities of the ephemera as form. With the word "Kalendar," he gestures toward another ancient form of annual record keeping: the gardener's almanac, a genre that includes Cato and Varro's *De re rustica*, John Evelyn's *Kalendarium Hortense*, and Gilbert White's *Natural History and Antiquities of Selborne*, the latter of which Thoreau knew well (see fig. 1.3).

Already, though, Thoreau knows that *his* Kalendar would be a highly personal document, not a mere table of statistics. "This may not be an annual phenomenon to you," he says of the appearance of the lodges, "but it has an important place in *my* Kalendar" (emphasis added). His Kalendar, when it began to take shape in March and April of 1860, would be a revelation: a new way of seeing the natural world and experiencing time.

Thoreau's habit from the early 1850s on was to spend the afternoon of every day walking, taking notes in pencil on small scraps of paper that he carried in his pockets. The next morning he would spend at his desk, elaborating lengthy and detailed descriptions from the previous day's scraps. The immediacy of the journal—which often slips into present tense, or adopts oddly timeless modes such as the future tense "you will hear" above—can make it easy to forget that it is not a real-time, in situ document. The entry above would have been written the following day, April 20, at his desk in the attic room of the family house on Main Street. Thoreau's daily practice thus entailed a complex two-part procedure: walking/observing, an activity that already encompassed the germ of his writing; and writing, a process that seemed to slip back into the experience of walking, and at the same time to anticipate the next day's walk (fig. 3.4).

As Laura Dassow Walls writes, Thoreau's journal practices reflect "an activity of the mind engaged in a reciprocal process: the mind enters nature, nature is taken into the mind; self and nature react on and finally make, and remake, each other."[5] In January of 1860 he'd written,

> A man receives only what he is ready to receive, whether physically or intellectually or morally, as animals conceive at certain seasons their kind only. We hear and apprehend only what we already half know. If there is something which does not concern me, which is out of my line, which by experience or by genius my attention is not drawn to, however novel and remarkable it may be, if it is spoken, we hear it not, if it is written, we read it not, or if we read it, it does not detain us. Every man thus tracks himself through life, in all his hearing and reading and observation and traveling. His observations make a chain. The phenomenon or fact that cannot in any wise be linked with the rest which he has observed, he does not observe. (J 13:77)

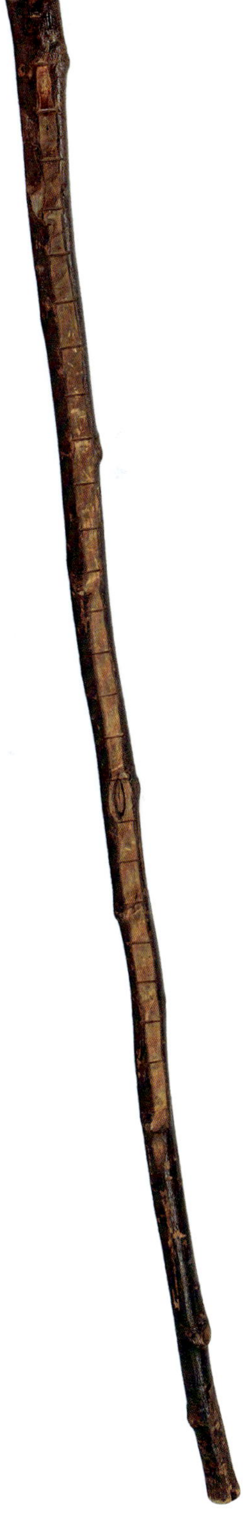

Thoreau's description of the way certain learning experiences prepare us for others serves as a gloss on his own practices. Each of his principal activities—walking, observing, and writing—helped prepare him for the others, so that these practices became a self-reinforcing chain. The description also reflects Thoreau's awareness of the complex temporal dimension of his process, the way he "tracks himself through life." We might read the "you" of the 1860 passage above ("you will hear at first a single beat or two far apart") as Thoreau talking to the self he is tracking, the future self who will hear again in some future April what he heard on April 19, 1860, and recorded the following day. The single observation of the partridge drum is linked both to earlier observations that prepared the way, and to future observations that he anticipates. By this period in Thoreau's life, he'd become adept at temporal thinking: he regularly stretched the present tense of writing backward to cover the previous day's experience. And he'd used something like the opposite strategy in *Walden,* condensing two years into one, making its shape into a single year that was also a single day, framed between chanticleer's morning call in the book's epigraph and the "morning star" of the last sentence.

These deliberate experiments in temporal reorganization had their origin in the world-shattering grief Thoreau experienced after the death of his brother, John, whose death had coincided with that of Ralph Waldo Emerson's beloved five-year-old son, Waldo. On the latter occasion, Thoreau had written to Emerson,

> Nature is not ruffled by the rudest blast. The hurricane only snaps a few live twigs in some nook of the forest. The snow attains its average depth each winter, and the chic-adee lisps the same notes. The old laws prevail in spite of pestilence and famine. No genius or virtue so rare and revolutionary appears in town or village, that the pine ceases to exude resin in the wood, or the beast or bird lays aside its habits.

> How plain that death is only the phenomenon of the individual or class. Nature does not recognize it, she finds her own again under new forms, without loss.[6]

From this time forward, Thoreau's response to loss involved immersing himself in seasonal time, within which even catastrophic loss could come to seem absorbable and temporary, finally insubstantial.

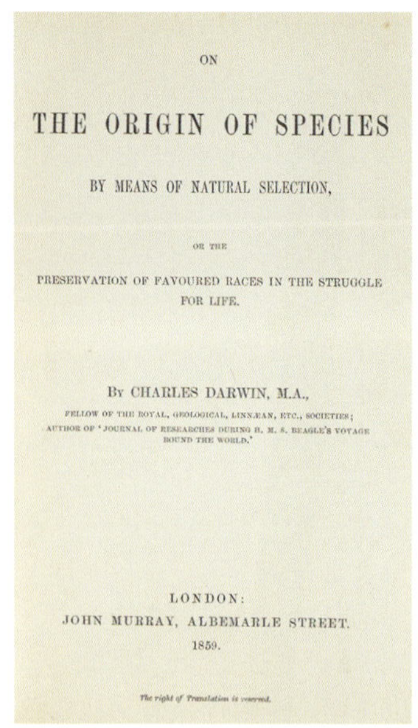
ON

THE ORIGIN OF SPECIES

BY MEANS OF NATURAL SELECTION,

OR THE

PRESERVATION OF FAVOURED RACES IN THE STRUGGLE FOR LIFE.

BY CHARLES DARWIN, M.A.,

FELLOW OF THE ROYAL, GEOLOGICAL, LINNÆAN, ETC., SOCIETIES;
AUTHOR OF 'JOURNAL OF RESEARCHES DURING H. M. S. BEAGLE'S VOYAGE ROUND THE WORLD.'

LONDON:
JOHN MURRAY, ALBEMARLE STREET.
1859.

The right of Translation is reserved.

Fig. 3.5. Charles Darwin, *On the Origin of Species by Means of Natural Selection; or, The Preservation of Favoured Races in the Struggle for Life* (London: John Murray, 1859). The Huntington, Burndy Library Collection, 712519.

In the spring of 1860, Thoreau was recovering from two perspective-shifting experiences: the raid on Harpers Ferry and John Brown's subsequent execution in late 1859, and his reading, on New Year's Day of 1860, of an advance copy of Charles Darwin's *On the Origin of Species* (fig. 3.5). Each of these events was, in its own way, a cataclysm. Brown's antislavery martyrdom redefined for Thoreau what it meant to live and act with integrity in relation to the slave state. Darwin illustrated in a new and comprehensive way the workings of more-than-human time, and the "complex web of relations" that composed the natural world.[7] In the wake of these encounters, Thoreau sought a wider vision of both time and relation.

Brown's violent insurrections in Kansas, the front lines of the conflict between pro- and antislavery factions after the passage of the Kansas-Nebraska Act in 1854, represented for Thoreau the actions of an uncompromising moral hero. This heroism, for Thoreau, was based on Brown's full recognition of the immorality of slavery and the implications of his relations to it. In "Resistance to Civil Government" Thoreau had written,

> If the injustice is part of the necessary friction of the machine of government, let it go, let it go: perchance it will wear smooth—certainly the machine will wear out. If the injustice has a spring, or a pulley, or a rope, or a crank, exclusively for itself, then perhaps you may consider whether the remedy will not be worse than the evil; but if it is of such a nature that it requires you to be the agent of injustice to another, then, I say, break the law. Let your life be a counter-friction to stop the machine.[8]

The Fugitive Slave Act conscripted every citizen of the United States into the service of injustice, and Thoreau saw in Brown's actions a divinely inspired counter-friction to the national machine.

In "A Plea for Captain John Brown," a speech first delivered in October 1859, two weeks after the raid on Harpers Ferry, Thoreau compared the nation to an overcrowded slave ship. The physical proximity suggested by this metaphor reflects his sense, heightened first through the passing of the Fugitive Slave Act in 1850 and underscored through his 1857 encounter with Brown, of the inescapable relation of every citizen of the United States to the reality of slavery.

> The slave-ship is on her way, crowded with its dying victims; new cargoes are being added in mid-ocean; a small crew of slaveholders, countenanced by a large body of passengers, is smothering four millions under the hatches, and yet the politician asserts that the only proper way by which deliverance is to be obtained, is by "the quiet diffusion of the sentiments of humanity," without any "outbreak." As if the sentiments of humanity were ever found unaccompanied by its deeds, and you could disperse them, all finished to order, the pure article, as easily as water with a watering-pot, and so lay the dust. What is that that I hear cast overboard? The bodies of the dead that have found deliverance. That is the way we are "diffusing" humanity, and its sentiments with it.[9]

In Thoreau's extended metaphor, there are no nonparticipants. The enslaved are the ship's cargo, directly oppressed by a small crew of slaveholders, who are "countenanced" by the "large body of passengers," who allow the violence to continue. In the logic of the metaphor, the solution is clearly a revolt against the crew on the part of the greater number of passengers. But action is deterred by politicians, who urge only "the quiet diffusion of the sentiments of humanity" and warn against "outbreak." As if to break the spell of euphemism, Thoreau literalizes this banal expression, recasting "diffusion" as the throwing overboard of the bodies of the enslaved.

When Thoreau encountered an early edition of *On the Origin of Species* on New Year's Day, 1860, just a few weeks after Brown's execution, he had thus already been grappling with the implications of seeing the world around him in terms of interdependent systems of relation. The atmosphere in Concord in these weeks also helps explain why at least initially, for Thoreau and his abolitionist friends, "*Origin of Species* was first and foremost an argument debunking the so-called scientific basis for slavery."[10] On January 1, Thoreau attended a dinner party at Franklin Benjamin Sanborn's house with Amos Bronson Alcott and Charles Loring Brace. Brace had recently arrived from a visit with his uncle, Harvard botany professor Asa Gray, an old friend of Darwin's, who had been sent an advance copy of *Origin*. As Walls writes, "Gray lent it to Brace, who showed it to Sanborn, Alcott, and Thoreau. All afternoon the four friends read [*On the*] *Origin of Species* aloud to one another and discussed Darwin's extraordinary principle of 'Natural Selection.'"[11] This theory overturned the picture promoted by the well-known biologist and Harvard professor Louis Agassiz of "spontaneous generation." According to the latter, unchanging individual species spontaneously appeared, uniquely (and divinely) formed. This theory, with its emphasis on an absolute and hierarchical distinction between species and races, formed the basis of the central "scientific" justification for slavery. Darwin confirmed Thoreau's conviction that Agassiz's picture was wrong. He also confirmed what Thoreau

		52	53	54	55	56	57	58	59	60
10	River when lowest in April & [illegible] — Ap 4-56 · not yet so high [illegible]	The high water year [illegible] May 15 [illegible]	[illegible]	23 [illegible] · 30 highest for April	May 4 [illegible] meads. 29 rapidly going down [illegible]	12 going down [illegible] May 1st water [illegible]	May 1st (?)	[illegible]	10th [illegible] · 17 up again [illegible]	lowest [illegible] the 31st [illegible] lowest Ap 30 [illegible]
16	River when highest in April — (May 31-50 higher than it has been at the season for many years — water over [illegible])	10 high but not high for spring. 23 highest for year 8½ in [illegible]	7 [illegible] great meads. Mar 17 [illegible] May 7 (?) [illegible] not high.	24 [illegible] little · 29 [illegible] May 5 [illegible] 23·52	14 steadily rising since [illegible] 19 fallen a little. 21 risen again a little [illegible] 23 at height for this [illegible] 28 [illegible]	11 at its height [illegible] 23 [illegible] 5 [illegible] 8 [illegible]	24 [illegible] higher than before. 22 higher [illegible]	14 a little higher [illegible] May 14 [illegible]	1st [illegible] 10 cannot [illegible] 28 [illegible]	[illegible] from Ap. 7 [illegible] falls from 15 [illegible] highest the 14 [illegible]
14	Smooth reflecting water	11 Placid [illegible] reflecting water		26 - overcast	15 still cloudy day water smooth & full of reflections	15 [illegible] overcast	May 3-57 [illegible]		17 [illegible]	
12	Ripples on ripple lakes									
13	Spearing	8 & 30	1st	5 & 8	16 - 25	25			9 [illegible] · 8 [illegible]	15 F. H. [illegible] · 26
Mar 25	1st leave off great coat (Mar.)	Mar 15 · [illegible] Ap. 26	Mar 26 · Ap. 13 [illegible]	Mar 17 [illegible] Ap. 25 [illegible]	16 - 1st [illegible] great coat 17-18-& 19 [illegible] 29 without great coat	7 & 8 [illegible]	Feb. 25 (54?) [illegible]	[illegible] 7 [illegible]	Mar. 17 [illegible] 27 & 30	Mar 31 1st [illegible]
15 (?)	Wear great coat again				28-29 & 30 · 55 · 10th			7		[illegible]
21	Begin to wear [illegible] coat [illegible]								[illegible] 21 (?)	19 - 60 [illegible]
27	Sit without fire [illegible]	30		26 [illegible]	19		May 3	May 1st 82	30	20 [illegible]
	1st [illegible]				19 · 55					
	Weather [illegible] coat								27 & 30	28 & 65 + [illegible]
20	Dark evening			26 - 54 [illegible] cloud.	17 rainy & very dark					16 - rainy & very dark
23	Burning of brush [illegible]			23	18 smoke in horizon					27
25	Fires in woods [illegible]	28							21 [illegible]	27
16	Plowing & planting going on generally	17 -			19	16-56		15 [illegible]	[illegible]	12
21	Showery clouds						21 - 24 - 25			
10	Dark based spring clouds	10					24			
18	[illegible] clouds hanging about — [illegible]				18					
21	Sea [illegible]				18 [illegible]	24 & 28-29 & 30				
7	[illegible]			[illegible] Mar 17	24 [illegible]	12 [illegible] 6 [illegible]		2		
17	Fog over river in morning				some 21st	28 [illegible] fog		6 [illegible] foggy [illegible]		
24	Cold in last half of month				27 & 28 [illegible]		28 [illegible] cold NW wind · 28 [illegible]	24 cold NW wind [illegible]		18 cold [illegible] 25-44+
25	[illegible] spotted with snow	1st & 4 & 28			18					
	1st moon to walk by			11 · 54						
19	Willow [illegible]		19 [illegible]	27 [illegible]		24 [illegible]				17 [illegible]
19 (?)	Bass [illegible]		19 [illegible]							
19 (?)	[illegible]									18 [illegible]
20 (?)	[illegible] aspen [illegible]									18 just begins
14	Rain after E wind	18 [illegible]							13-14 [illegible]	11 [illegible]
21	White frosts	[illegible]		12 (after clear moon-light)	21 [illegible] 8 [illegible]					29 [illegible] 30 [illegible]
17	Evenings very [illegible]									17
25	[illegible] leaves [illegible] young oaks									26
	Begin to sit without fire [illegible]	30		26	19		May 3	May 1st 82	30	20

had been learning from Brown's example: that within a given system, seemingly disparate individuals—say, an enslaved person in a southern state and a northern White abolitionist—were, in reality, connected. As Darwin put it, even "plants and animals most remote in the scale of nature, are bound together by a web of complex relations."[12]

In the months that followed, Thoreau would grapple with the implications of natural selection and its picture of self-governing natural systems in his lecture "An Address on the Succession of Forest Trees." Printed in the *New-York Weekly Tribune* and widely reprinted, this text was read by more people in his lifetime than anything Thoreau had ever written, and would go on to become "a pioneering document in the development of forest management and of ecology."[13] "Succession" describes the development of a forest ecosystem through the interwoven agency of plants, animals, and environmental factors. Trees appeared in new areas not through divine intervention, but through the agency of seeds shaped by natural selection for distribution by wind or via the bodies of animals and birds. While some seeds fell on lakes or rocks, others found fertile soil and flourished. The growth of a single tree thus depended on the interaction of many factors within a given landscape—just as the cruel progress of the slave ship described in "A Plea for Captain John Brown" was made possible by the interactions of crew members, politicians, and passengers.

It was in the dizzying, perception-altering light of these experiences of interdependence and systems-thinking that Thoreau created the first charts of general phenomena in March and April of 1860 (fig. 3.6). These charts, drawn up on surveying paper, compile information about the weather-related phenomena of the month from the previous ten years of Thoreau's journal. Along the left-hand side of the page, Thoreau listed the phenomena that seemed to him most typical of the month. As always, Thoreau chose his words deliberately: the word "phenomena" reflects a post-Kantian understanding of the natural world as mediated by human perception, and Thoreau's categories frequently betray an equal interest in observing "subject" and observed "object"—indeed, they bring this epistemological separation into question. Unlike the charts of bird migration, leaf-fall, and spring flowering times, which are organized by date and species name with little in the way of "subjective" observation, the monthly charts record categories such as "1st leave off greatcoat," and "sit without fire commonly."[14] Running across the top of the chart, Thoreau created columns for each year, beginning in the early 1850s and ending in 1860 or 1861. In the grid created by these two axes, Thoreau copied information from the journal—usually dates accompanied by descriptive words or phrases.

One of the categories recorded in April of 1860 is "sea turn"—a cool wind coming from the ocean and creating a sudden change in the atmosphere—which

Fig. 3.6. Henry David Thoreau, "General Phenomena for April," autograph manuscript, c. 1851–60. The Morgan Library & Museum, New York, purchased by Pierpont Morgan with the Wakeman Collection, 1909, MA 610.

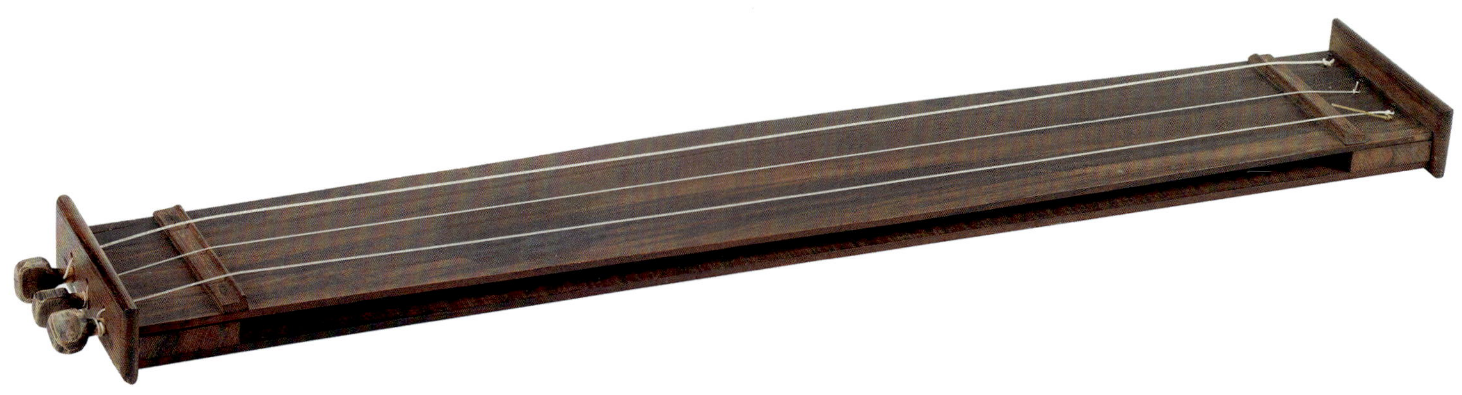

Fig. 3.7. Henry David Thoreau, aeolian harp, c. 1850–60, Concord, Massachusetts. Rosewood, 1½ × 4½ × 28¾ in. (3.8 × 11.4 × 73 cm). Courtesy of the Concord Museum, Gift of Mrs. Leon Foss, Th68.

keys to entries from 1855 and 1856. The 1855 entry refers to the following journal passage:

> When I reach the top of the hill I see suddenly all the southern horizon (east or south from Bear Hill in Waltham to the river) full of a mist, like a dust, already concealing the Lincoln hills and producing distinct wreaths of vapor, the rest of the horizon being clear. Evidently a sea-turn—a wind from over the sea, condensing the moisture in our warm atmosphere and putting another aspect on the face of things. All this I see and say long before I see the change, while still sweltering on the rocks, for the heat was oppressive. Nature cannot abide this sudden heat, but calls for her fan. In ten minutes I hear a susurrus in the shrub oak leaves at a distance, and soon an agreeable fresh air washes these warm rocks, and some mist surrounds me. (J 7:321)

In the journal, Thoreau describes the external phenomenon of the sea turn in terms of his visual, auditory, and haptic perception. Watching the condensed air approach, he notes that he can see and describe it ("all this I see and say") before he directly experiences the cooler atmosphere. Remaining in the present tense, he notes "in ten minutes I hear a susurrus in the shrub oak leaves"—this then is the interval between his initial visual perception and the more immediate signal that the ocean air has arrived: a rustling sound in the leaves around him. At last, he is enveloped in the cool air: "some mist surrounds me."

Given Thoreau's long-standing interest in the effects of atmosphere, it is not surprising that "sea turn," despite its relatively rare occurrences in Thoreau's journal, made it onto the list of general phenomena for April, where it appears alongside other atmospheric categories such as "hazy" and "fog over river in morning." Linking the external world to internal experience, "atmosphere," like "phenomenon," mediates between subject and object. Interest in atmosphere and atmospheric effects was growing in the nineteenth century, in part due to increased knowledge of atmosphere as a material entity: landscape painters, notably J. M. W. Turner and John Constable, emphasized the sky as active sources of light and meaning in their paintings. Constable's 1821–22 studies of cloudscapes, possibly inspired by the theories of meteorologist Luke Howard, illustrate the way atmosphere, once considered as a background to landscape, was becoming a subject in its own right (see images on pages 114 and 115).[15]

Thoreau's interest in atmosphere and atmospheric effects is also illustrated by his fascination with sound produced by the wind: he owned an aeolian harp (fig. 3.7)—a stringed instrument sounded by the wind—and commemorated its haunting music in a poem ("Rumors from an Aeolian Harp") in *A Week on the Concord and Merrimack Rivers*.[16] A visible, audible, and material occurrence in the world, atmosphere enters the body of the observer through the multiple channels of sense perception and through the breath. Surrounding us, it has no definite location, but is a defuse presence, a whole environment. In *Walden* Thoreau had written,

> The gentle rain which waters my beans and keeps me in the house today is not drear and melancholy, but good for me too. Though it prevents my hoeing them, it is of far more worth than my hoeing. If it should continue so long as to cause the seeds to rot in the ground and destroy the potatoes in the low lands, it would still be good for the grass on the uplands, and, being good for the grass, it would be good for me. . . . In the midst of a gentle rain while these thoughts prevailed, I was suddenly sensible of such sweet and beneficent society in Nature, in the very pattering of the drops, and in every sound and sight around my house, an infinite and unaccountable friendliness all at once like an atmosphere sustaining me, as made the fancied advantages of human neighborhood insignificant, and I have never thought of them since. Every little pine needle expanded and swelled with sympathy and befriended me.[17]

The passage identifies the causes of this atmospheric experience as both external and internal, objective and subjective: the experience occurs "in the midst of a gentle rain" and also "while these thoughts prevailed." The rain and Thoreau's thinking about its value to the more-than-human world conspire to create an atmosphere of "infinite and unaccountable friendliness." This affectively warm and beneficent atmosphere seems in turn to heighten Thoreau's sensory perception, to open his senses to "the very pattering of the drops" and to "every little pine needle." Taken as a whole, the passage describes the difference atmosphere makes to both feeling and perception, and how these in turn affect the atmospheric phenomenon.

Thoreau observed, noted, and sought again the conditions in which atmospheric effects were most readily perceived. He particularly delighted in the phenomena of the *echo*, in which the environment itself becomes a resonant instrument:

> All sound heard at the greatest possible distance produces one and the same effect, a vibration of the universal lyre, just as the intervening atmosphere makes a distant ridge of earth interesting to our eyes by the azure tint it imparts to it. There came to me in this case a melody which the air had strained, and which had conversed with every leaf and needle of the wood, that portion of the sound which the elements had taken up and modulated and echoed from vale to vale.[18]

In Thoreau's description, distance is not primarily a subtractive phenomenon, representing absence, but an additive one, positively contributing to the phenomenon experienced. Sound, in traveling to us from a distance, is not *less* (though it is likely quieter) than nearby sound—in fact, it carries with it the sonic traces of the distance it has traveled. Similarly, in Thoreau's formulation, distance "imparts" a new color to the ridge, adding to its visual effect. Far from seeking to minimize what others might describe as atmospheric *distortion*, Thoreau reveled in experiences that allowed his senses to collaborate with the effects of the external world. The blending of internal and external, subjective and objective experience that characterize Thoreau's description of individual phenomena such as "sea turn" also reflect the orientation of the Kalendar as a whole, which charted the months not only as they happened to him, but also as he participated in their unfolding.

If the Kalendar mediates between subjective and objective perception, it also mediates between temporal modes. On the one hand, the charts of general phenomena provide a comprehensive overview of the given months during the span of a decade. However, they also provide an index of particular, extended moments in time, moments Thoreau preserves through detailed description in journal entries. Taken together, the Kalendar and the journal function as a unique perceptual apparatus, enabling Thoreau both to observe overall patterns and also to "zoom in" on individual instances. Having invented this technology for seeing the seasons of his life in the spring of 1860, Thoreau seems to have dropped the project in the fall of that year. He would pick it up again a year later, in a very different season of his life.

Part II: Fall 1861

> It excites me to see early in the spring that black artery leaping once more through the snow-clad town. All is tumult and life there, not to mention the rails and cranberries that are drifting in it. Where this artery is shallowest, i.e., comes nearest to the surface and runs swiftest, there it shows itself soonest and may see its pulse beat. There are the wrists, temples of the earth. Where I feel its pulse with my eye. The living waters, not the dead earth. It is as if the dormant earth opened its dark and liquid eye upon us. (*J* 13:138)

Thoreau's interest in the Concord River (and its tributaries, the Assabet and Sudbury rivers), was long and deep. The Sudbury was visible from his family home on Main Street, and it was from his nearby "boat place," also on the Sudbury, that he headed out on excursions throughout the 1850s, often accompanied by his friend and neighbor William Ellery Channing. In many of the entries of these years, Thoreau's famous *I* is replaced by a *we*. In November of 1853, he writes,

> It is a full moon—& a clear night—with a strong northwest wind—so C and I must have a sail by moonlight. The river has risen surprisingly to a spring height owing to yesterday's rain—higher than before since spring—We sail rapidly upward—The river apparently almost actually as broad as the Hudson—Venus remarkably bright just ready to set—not a cloud in the sky—only the moon & a few faint unobtrusive stars here & there & from time to time a meteor— . . . It is very pleasant to make our way thus rapidly—but mysteriously over the black waves—now black as ink & dotted with round foam spots with a long moonlight sheen on one side—to make one's way upward thus over the waste of waters not knowing where you are exactly only avoiding shores—The stars are few and faint in this bright light—How well they wear. C. thought a man could still get along with *them* who was considerably reduced in his circumstances—that they were a kind of bread & cheese—that never failed. (*PJ* 7:162)

The entry gives us a sense of what Thoreau and Channing shared. For both men, a full moon, a clear night, and a strong wind amount to an imperative to be out on the water. And for both, connection to the natural world was sustenance, "a kind of bread & cheese—that never failed."

But this sustenance had become scarce for Thoreau by the fall of 1861. Having contracted bronchitis (worsened by underlying tuberculosis), he'd begun to fail the previous spring and had undertaken a journey to Minnesota in an

General Phenomena for October

	52	53	54	55	56	57	58	59	60	61
River [illegible] [illegible]	[illegible]	[illegible]	[illegible]	Sept. 24 [illegible]	[illegible]	[illegible]	[illegible]	[illegible]	[illegible]	[illegible]
River Highest										25 [illegible]
Rain in 1st half [illegible]	10 [illegible] 5 night & morning	[illegible]		[illegible]	Oct 1 [illegible]	15 [illegible]	13 [illegible]	2 in night	[illegible]	5 [illegible] 11
Rain in last half [illegible]	[illegible]	22 [illegible] 28 [illegible]	31st		18 [illegible]		30 rain with wind	18 rain [illegible]	22 [illegible]	[illegible] 19
NE storm [illegible]						25-26-[illegible]	24 NE storm [illegible]			
Rain filling springs & streams		24 rain all day filling the streams				24 [illegible]				19 (also)
Freshet		31st				31				
[illegible]						24 [illegible]				10 [illegible]
Thunder shower					6 [illegible]				26 [illegible]	
Lightning at eve										5 & 19
Springs low [illegible]						14 [illegible]				18 [illegible]
Farmers [illegible] Oct 8 51				18 digging [illegible]		9 [illegible] 11 [illegible]				22
Smoke in horizon					[illegible] 21	10 [illegible] horizon				
	52	53	54	55	56	57	58	59	60	
Walden [illegible]	[illegible]	23 [illegible]	[illegible]				26 [illegible]	The cold [illegible]		17 [illegible]
1st cold weather					Oct 1 [illegible]			Oct 3 [illegible]		Oct 9 [illegible]
Last of [illegible]		Oct 4 [illegible] 31 [illegible]	Sept 24 [illegible] 90.		27 about [illegible]		Sept 30 [illegible]			
1st decided coolness						17 cooler	Oct 25 [illegible]			13 [illegible]
Wear a thicker coat [illegible]					Oct 1 [illegible]	Before Oct 5	26 [illegible]			
[illegible] cold early or late					14 [illegible]		26 [illegible]	16 [illegible] 20 [illegible]	Oct 1 [illegible]	
Cold before 2d. [illegible]	15 [illegible]							10 [illegible]	[illegible]	
Cold after do	22 [illegible]	Oct 25 [illegible] 31 [illegible]		Oct 24 25 [illegible]		20 cool 21 [illegible]	21 cooler [illegible]	15 [illegible] 16 [illegible] 20 [illegible] 21 [illegible]		
Very cold									Oct 1 [illegible] 21 [illegible]	
Hard Frost Oct 17.51 [illegible]	[illegible]	Sept 11 [illegible] Oct 15 [illegible] 30 [illegible]	Sept [illegible]	Sept 20 1st decisive [illegible]	Oct 6 [illegible]	Sept 30 [illegible] Oct 10 & 11 [illegible] 22 [illegible]	Sept 25 & 26 [illegible] 29 hard frost [illegible]	Sept 15 & 16 [illegible]	Sept 10 [illegible] 29 & 30 [illegible]	[illegible] 25 [illegible]
Ice in [illegible] Sept 15, 51 [illegible]	15 [illegible]	Oct 15 [illegible]			15 [illegible]	21 [illegible]		17 [illegible]	Sept 28 [illegible]	
Ice along river side		31 a little [illegible]				21 [illegible]		21 [illegible]		
Fair at eve Oct 8 51 [illegible]						Before Oct 5				
Begin [illegible]					20 [illegible]	17 [illegible]				
1st snow Oct 27 51 [illegible]	15 [illegible]	Nov. 8	Nov 15 [illegible]	Nov. 17 1st snow		20 1st snow [illegible]		Nov 12 [illegible]		
Snow on mts			19 [illegible]				Nov 7 [illegible]	20 [illegible] N.H.		

90

unsuccessful effort to improve his health. But by October, there could be little hope of recovery. Too weak for the daily walks that once provided the material for his journal, Thoreau largely abandoned it, writing his last entry in early November. Interestingly, at this point Thoreau revives the Kalendar project, drawing up charts of general phenomena in October, November, and December. Several of these charts contain particularly full entries on the height of the river, including precise measurements Thoreau was in no shape to have gathered for himself.

At the end of a list entry in early November, Thoreau notes a seasonal anomaly: "C. saw a strawberry in flower yesterday."[19] In these final months, Channing was a regular visitor at Thoreau's bedside, and he provided more than human company: he brought news of the woods and its unexpected flowerings. And perhaps more importantly, he brought news of the river.

Back in August 1852, Thoreau had noted in his journal, "The river is 8 1/12 feet below top of truss add 8½ [inches] for its greatest height this year you have 8 feet 9½ inches for the difference. It is ap. as low now as the first week in July. . . . Those are the limits of our river's expansibility—so much may it swell" (PJ 5:307).

In the October chart, Thoreau refers back to this calculation, adding later data and calculating again, as though he sought to sense the river in a new way: not by placing himself physically in its current, but by mentally tracking its seasonal ebb and flow (fig. 3.8). Although—or perhaps because—his usual habits of observation and recording had been disrupted, Thoreau threw himself into the project of re-creating "the limits of our river's expansibility."

Channing's involvement in this project was nothing new. In the summer of 1859, Thoreau, known for his meticulous work as a surveyor, as well as his knowledge of the river, was hired by a group of farmers engaged in a lawsuit against the owners of the Billerica dam, which, according to the farmers, was causing ruinous annual floods. Thoreau was given the assignment to "measure and catalog the width of each bridge that crossed the Concord between Sudbury and Billerica," as well as "the characteristics of every pier that might jut its obstructing pylons into the water, the history of each bridge's construction and improvement, and the character of the falls at Billerica itself."[20] Thoreau never ended up testifying: Robert Thorson suggests that the evidence didn't clearly support either side in the case. But in the course of data-gathering, he and Channing had undertaken extensive measurements of the river's height at various sites.[21] When his work for the farmers was concluded, he undertook a project of his own, transforming the data he and Channing had painstakingly gathered into a unique, large-scale survey. Studying the 1834 map of the river drawn by B. F. Perham under the supervision of the surveyor Loammi Baldwin, Thoreau both admired its clarity and noted its faults: not only its flat-out errors (mistakes

Fig. 3.8. Henry David Thoreau, "General Phenomena for October," autograph manuscript, c. 1852–61. Courtesy of Yale Collection of American Literature, Beinecke Rare Book and Manuscript Library, Henry David Thoreau Collection, YCAL MSS 203.

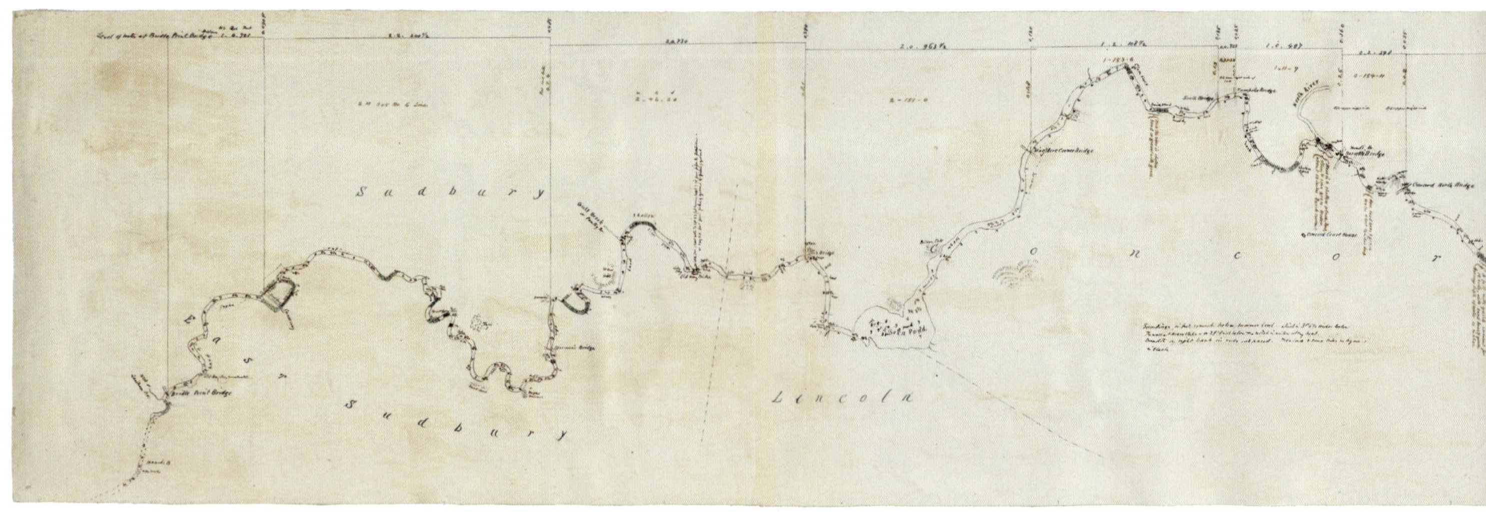

Fig. 3.9. Henry David Thoreau, *Plan of Concord River from East Sudbury to Billerica Mills, 22.15 Miles*, 1859, based on B. F. Perham's 1834 survey. Ink and graphite on paper mounted on cloth, 15 × 91 in. (38.1 × 231.1 cm). Concord Free Public Library, William Munroe Special Collections.

in the names of local bridges, inconsistencies in the measurements of the river's rise and fall), but also its lifelessness. Thoreau decided to correct the record by making his own (fig. 3.9). As Daegan Miller writes,

> Thoreau began by tracing the outlines of Baldwin and Perham's map and then reusing it for his own. But the two maps' similarities end at the sketched-in river's banks, for Thoreau's is alive with a riot of thousands of tiny notations—most in variously hued pen's ink, but some, ghostly, in pencil—including particular comments about the river's current: "shallow and quick" in some places, "sandbars and grass," "soft banks," in others. And there's a sense of the river's human use. Thoreau notes, just downstream from the Turnpike bridge, "1st cottage," and just a little below, the "boat pl.," from where he and his friend Channing began their riverine explorations. Elsewhere he marks the good swimming holes. He points out the cultural and historical geography of the river: where an old hay bridge once stood, where one might find freshwater clams, the monument to the battles of Lexington and Concord in 1776. And he carefully details, in scores of places, what plants grow where: individual oak and ash trees, polygonum, bulrush, and many others.[22]

As a tool allowing for both zoomed in and zoomed out perspectives, the river survey bears an important affinity to the Kalendar project. Both late-life projects unite Thoreau's desire for comprehensive perception of the natural world with his equally strong desire to record and preserve its particulars. Thoreau loved both skating and boating because they offered opportunities for perception-expansion. In February of 1855, he'd written, "I still recur in my mind to that

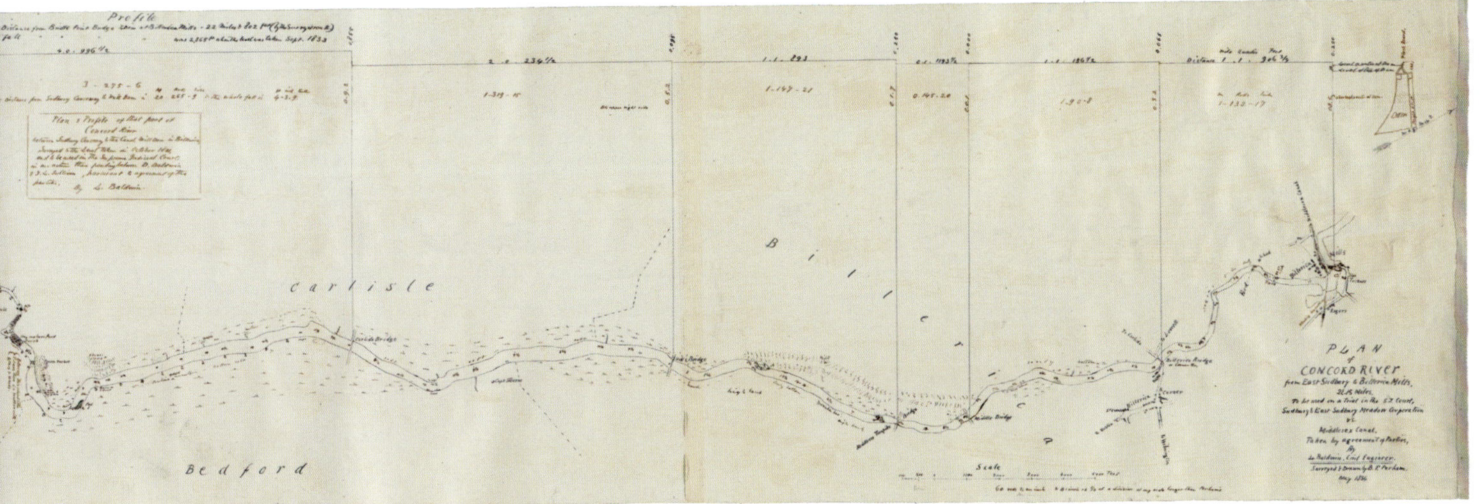

skate of the 31st. I was thus enabled to get a birds-eye view of the river—to survey its length and breadth within a few hours, connect one part (one shore) with another in my mind, and realize what was going on upon it from end to end,—to know the whole as I ordinarily knew a few miles of it only" (PJ 7:168). This desire for knowledge of the *whole* river, like his desire "to know all the phenomena of the spring"—"the entire poem" of the season—partakes of the nineteenth-century dream of comprehensive knowledge. As H. Daniel Peck writes, "The unstated assumption of the journal is that at some future point, a point always receding before the endless work of observation, the picture will be complete. At that moment, the master perceiver will collect his views and integrate them into a coherent vision, which would be nothing less than the world as seen in the mind of God."[23] And while it's possible to understand both of the charts of general phenomena and the river survey as partaking of this dream of mastery, both projects also reflect Thoreau's awareness of—in fact, his insistence on—the particularity of his own perspective. In fact, the map's most personal note, the element most indicative of Thoreau's particular life in connection with the great river and its tributaries, is also the baseline for its quantitative measurements. In his key to the survey, Thoreau writes,

> Soundings, in feet, so much below summer level—which is 3 ft 6½ inches below The wall at Hoar's steps—or 2 ft + 8 inch. Below the notch in my willow at my boat. Breadth on right bank in rods, as paced. Meadows & some hills &c. by me, in black.

"My willow at my boat" refers to a notched willow tree near where Thoreau kept his boat, and most likely, a stick for measuring the distance between the notch and the surface of the water. It was located in Channing's backyard.

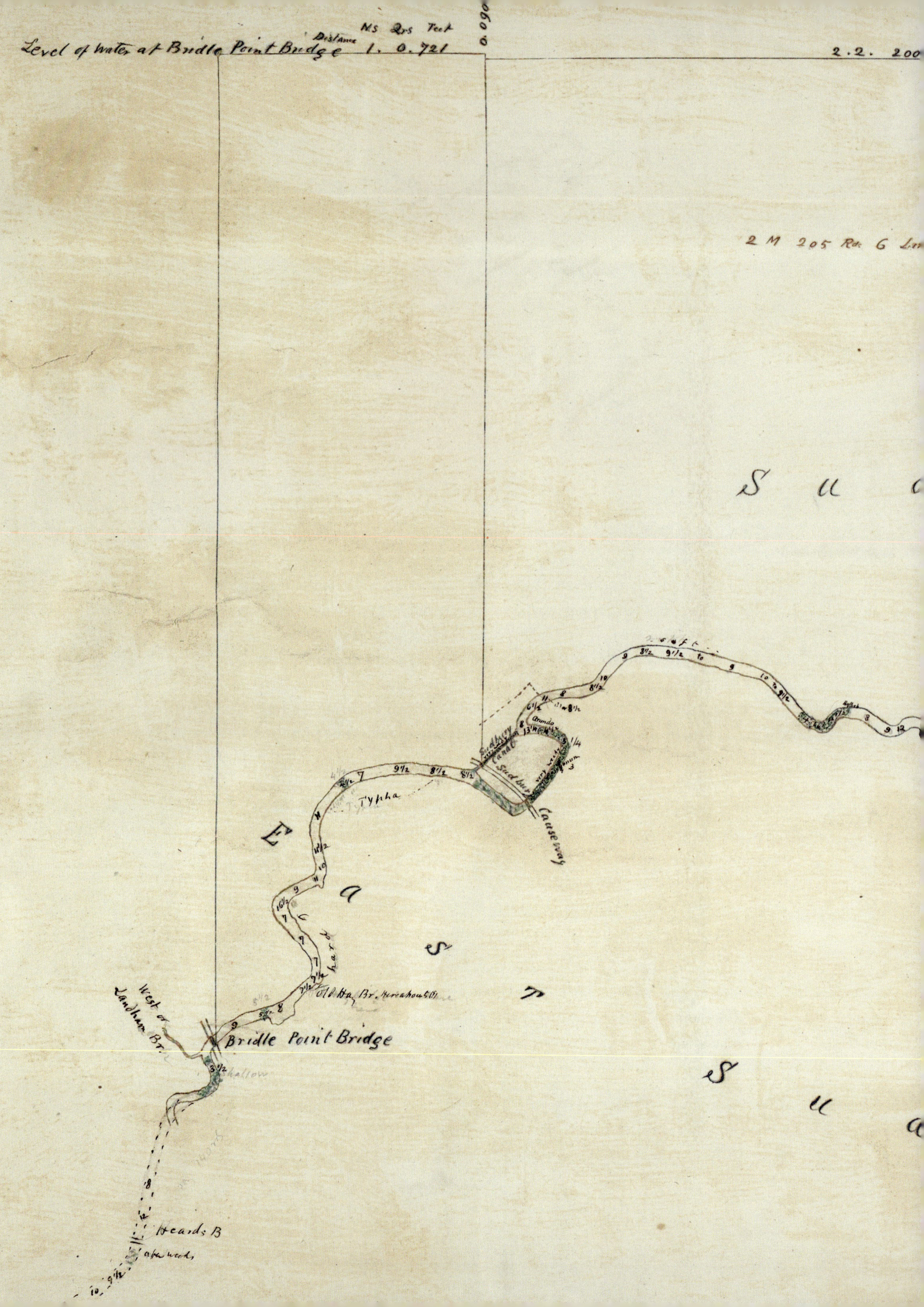

Level of water at Bridle Point Bridge

Distance

Ms 2rs Feet

1. 0. 721

0.090

2.2. 200

2 M 205 Rds 6 Lin

S u a

E a s t

S u a

Sudbury Canal

Causeway

Typha

Bridle Point Bridge

West of Landham Br.

Shallow

Heards B

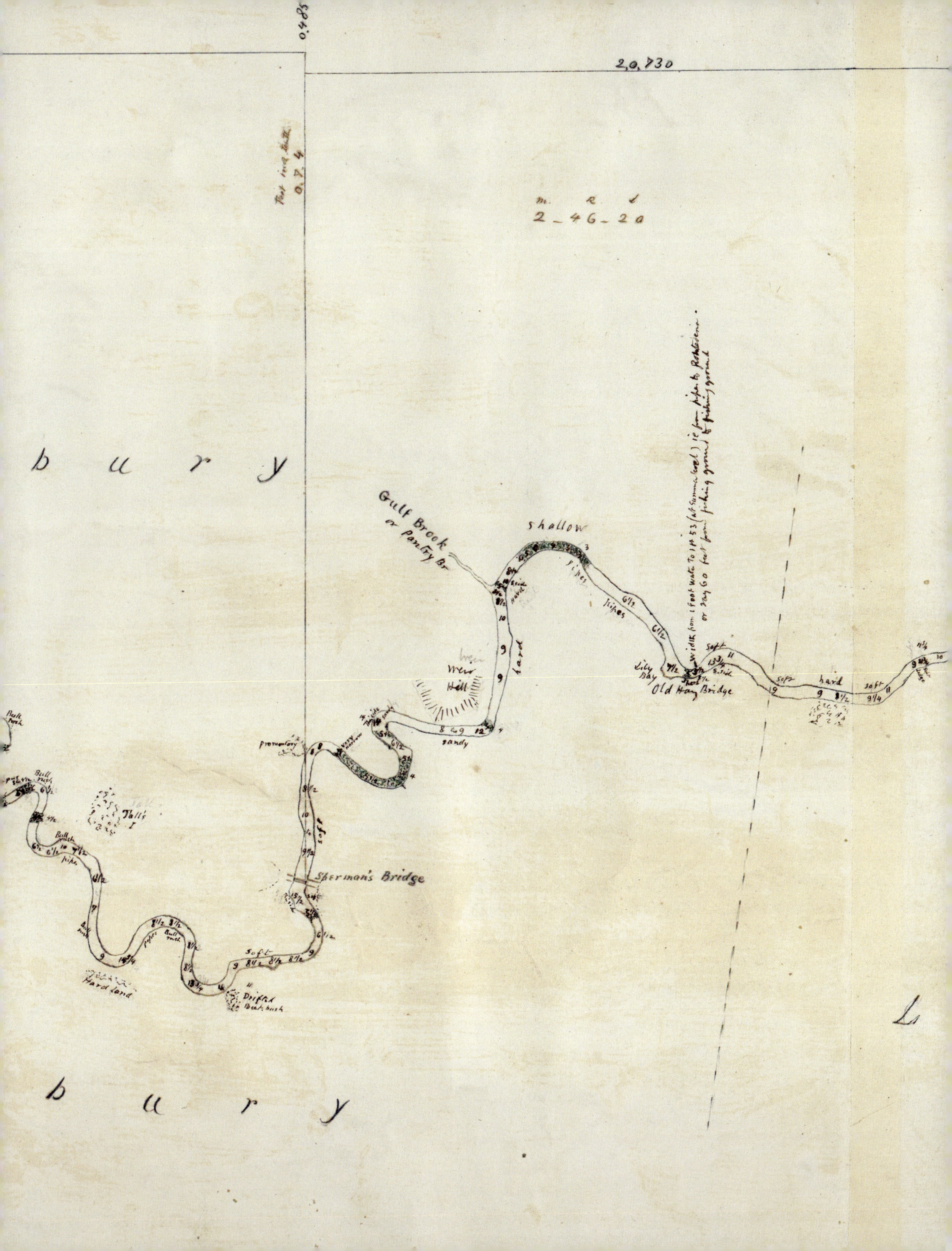

0.485

20,730

Test incq. Cell
0.7.4

m. r. l.
2_46_20

b u r y

Gulf Brook
or Pantry Br

shallow

width from 1 foot water to 1 ft 53 (at summer level) ie from pipes to Pontederia
or say 60 feet from fishing ground to fishing ground.

pipes

Weir Hill

hard

sandy

promontory

Lily Bay

Old Hay Bridge

soft

hard

soft

Sherman's Bridge

Tall's I

Hard land

Drifted Bull-bush

b u r y

L

0.390

2.0. 963 1/2

0.120

1.2. 108 1/2

1-153-6

0.8.1

2-151-0

0.1.8

Clam Shell

Here the river is shallow. Sand or gravel with grass.

Nine Acre Corner Bridge

mossy

9 1/2 to 10 muddy

Billern Cliff

Tav.

Lee's Bridge

O

N

muddy

Pole Brook

Fairhaven Pond

mud

Sound...

The wa...

Bread...

n'ble...

ncoln

0.135
0.025
0.360
0.035

0.0.723
1.0.497
0.2.398

0.31-22
1-11-7
0-154-11

0.1.9
0.3.6
0.0.2

B.l. lower right side of both

Turnpike Bridge

South Bridge

North River

Bl. upper right side

Bl. upper right side

Hay Ford

Hunt's Br
Barrett's Bridge

Shallow & quick

Here is a shallow place, sand, stones & some grass. Water about 13 inches. Quick current.

Here sand bars & grass. Water about 14 inches deep.

Rocks & Duck

monument

Concord North Bridge

Swim. Place

Oakes

Ash tree hole

Concord Court House

C O r

Here the river is 10½ rods wide. A fall with quick current for 15 rods, with sand bars & grass. Average depth of water 12 inches.

in feet, somewhat below summer level – which is 3 ft 6½ inches below
...ars steps – or 2 ft 8 in. below the notch in willow at my boat.
right bank in rods, as paced. Meadows & some Hills ... by me,

Whole Distance from Bridle Point Bridge to Dam at Billerica Mills = 22 miles & 802 feet (by the Surveyed rou
Whole fall " " " " was 2,865 ft when the level was taken Sept. 183

4.0. 996 1/2

3 – 275 – 6

	M	Rods	links		ft	and tenths
The Whole Distance from Sudbury Causeway to Mill Dam is	20	265	9	& the whole fall is	4	3.9

Plan & Profile of that part of
Concord River
between Sudbury Causeway & the Canal Mill Dam in Billerica,
Surveyed & the Level taken in October 1811,
and to be used in The Supreme Judicial Court,
in an action there pending between D. Baldwin
& J. L. Sullivan, pursuant to agreement of the
parties.
By L. Baldwin.

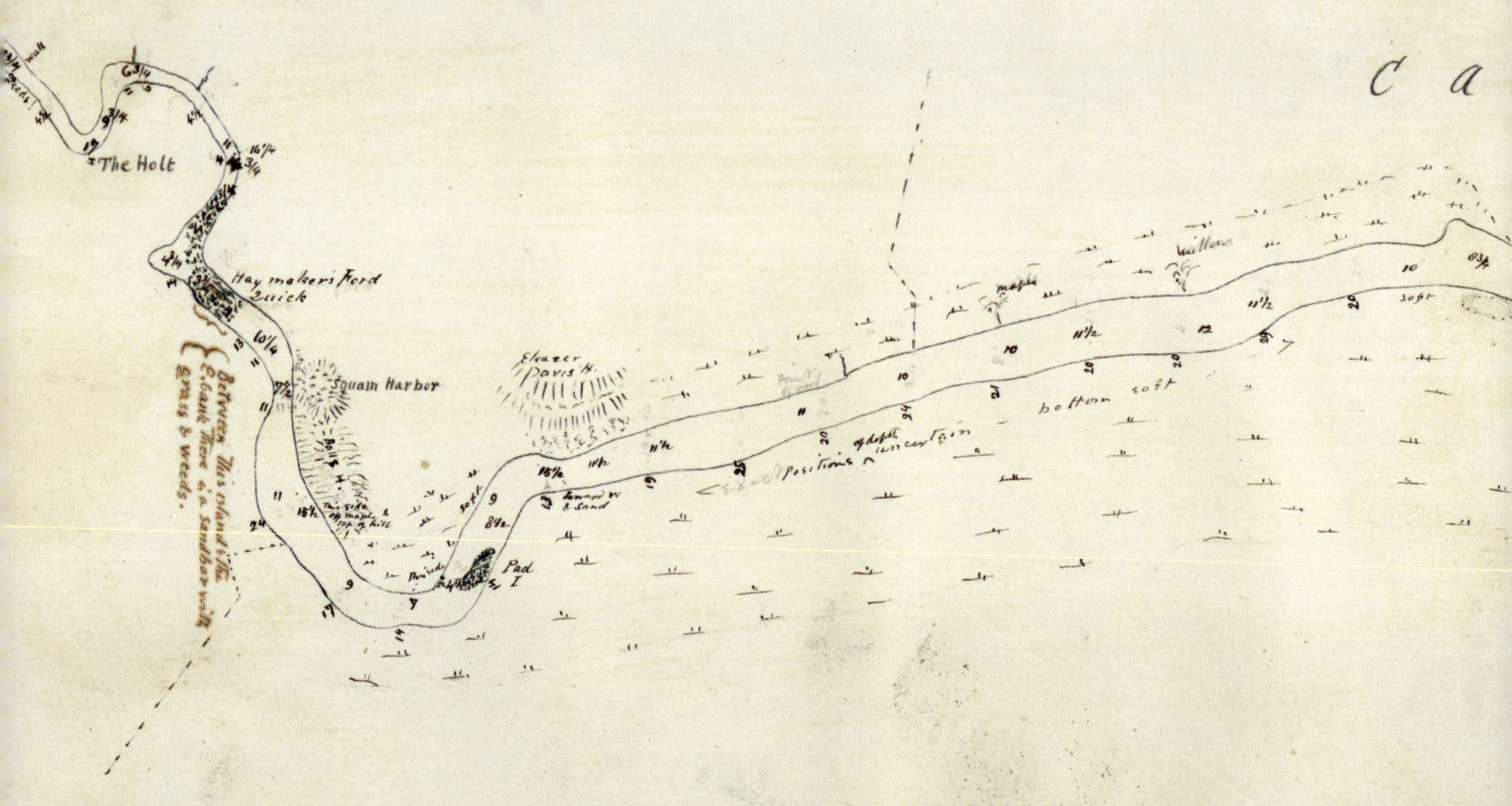

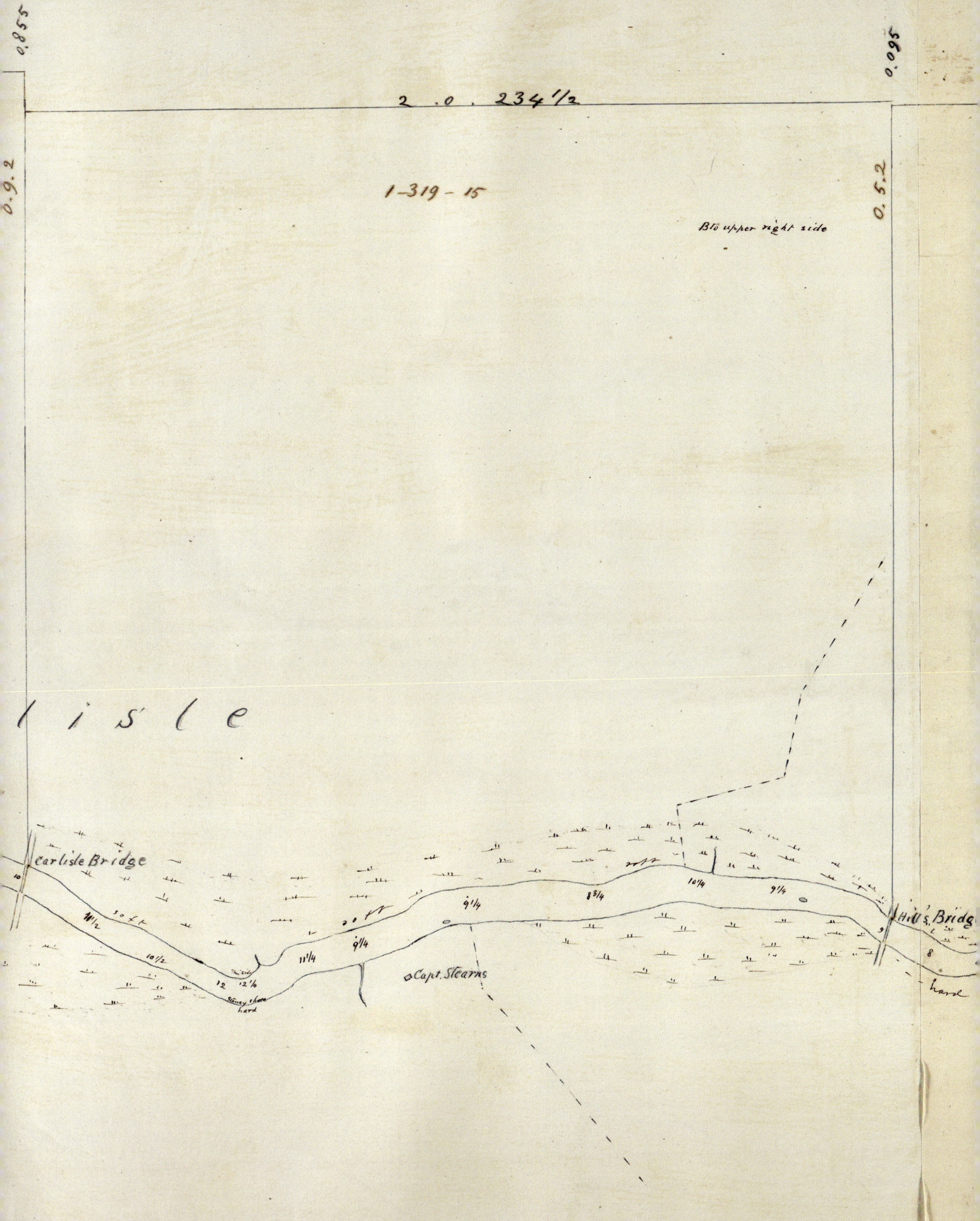

0.855

2 . 0 . 234 1/2

0.095

0.9.2

1-319-15

0.5.2

Blö upper right side

lisle

Carlisle Bridge

10

soft

11 1/2

10 1/2

12

12 1/4

Stoney shore hard

11 1/4

soft

9 1/4

9 1/4

Capt. Stearns

8 3/4

soft

10 1/4

9 1/4

Hill's Bridge

9

8

hard

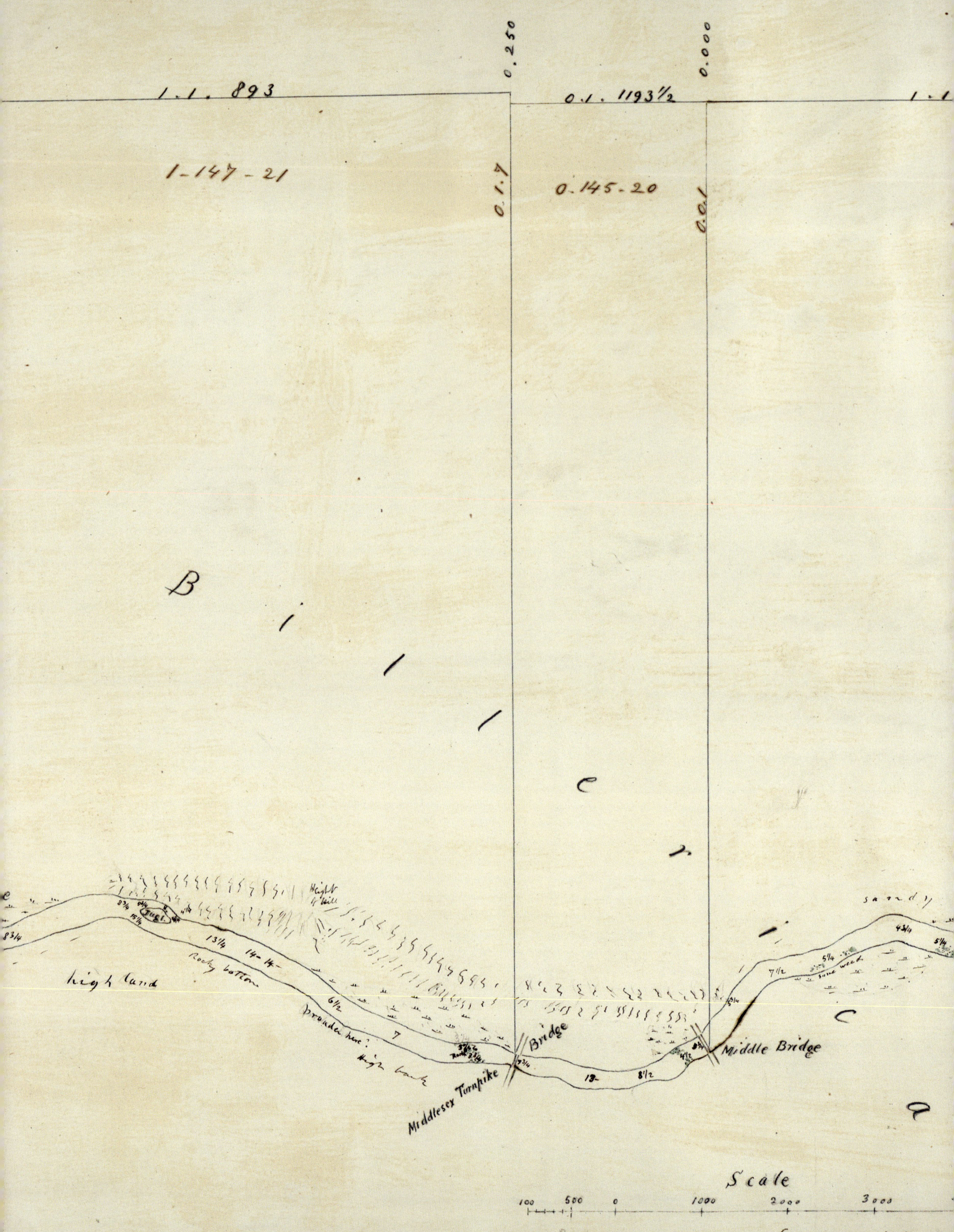
1.1. 893
0.250
0.1. 1193½
0.000
1.1
1-147-21
0.1.7
0.145-20
0.0.1
B i l l e r i c a
Hight of hill
13¼
14 14
Rocky bottom
high land
6½
Broader here?
7
High bank
Middlesex Turnpike
Bridge
18
8½
Middle Bridge
7½
some weeds
sandy
4¾
5¼
Scale
100 500 0 1000 2000 3000

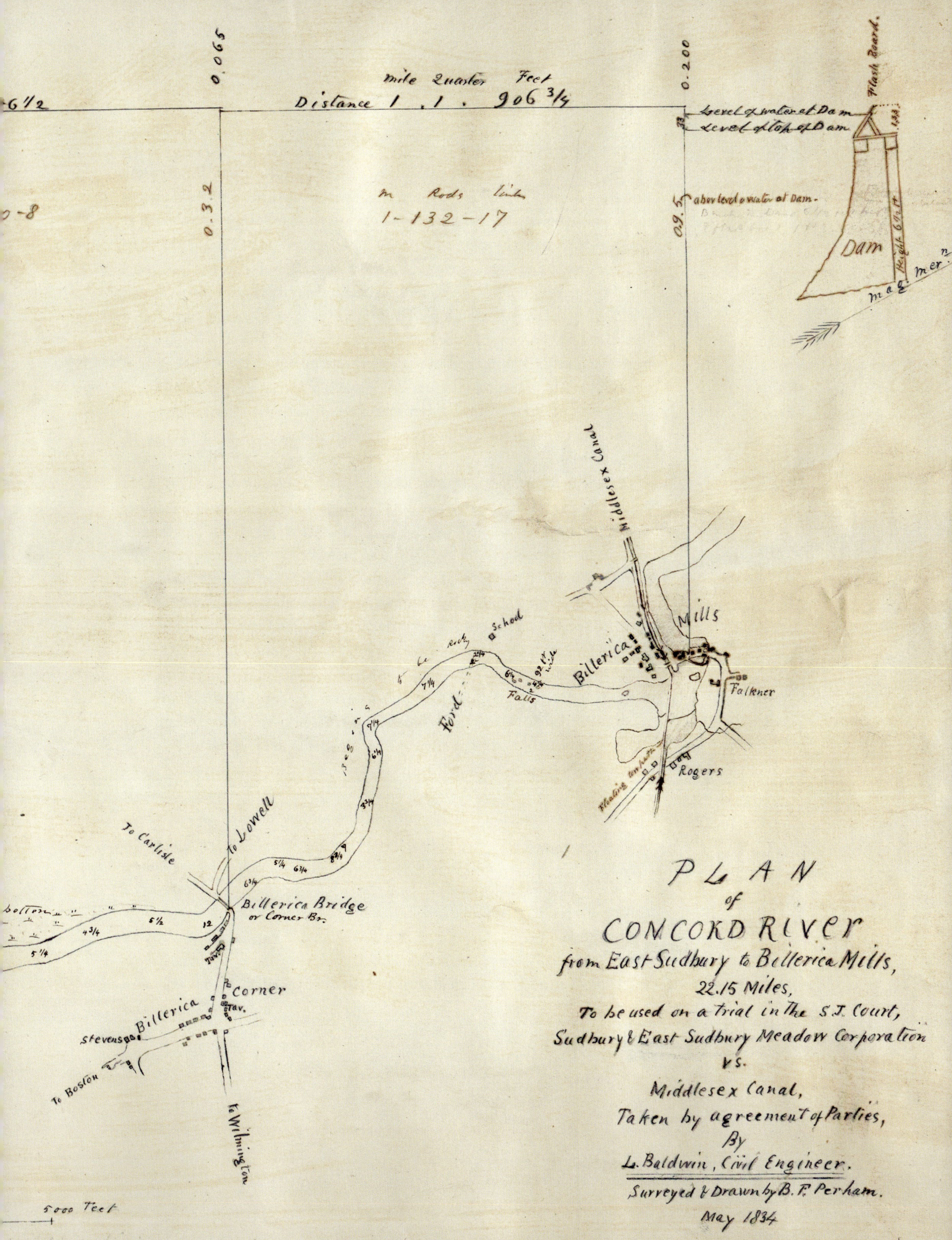

Mile Quarter Feet
Distance 1 . 1 . 906 3/4
0.065
0.200
0.3.2
m Rods link
1-132-17
Level of water at Dam
Level of top of Dam
Flash Board.
above level of water at Dam
Dam
Mag. Mer.n
Middlesex Canal
Mills
School
Billerica
Falls
Ford
Falkner
Rogers
To Carlisle
to Lowell
Billerica Bridge
or Corner Br.
Corner
Tav.
Billerica
Stevens
To Boston
to Wilmington
5000 Feet
PLAN
of
CONCORD RIVER
from East Sudbury to Billerica Mills,
22.15 Miles,
To be used on a trial in the S.J. Court,
Sudbury & East Sudbury Meadow Corporation
vs.
Middlesex Canal,
Taken by agreement of Parties,
By
L. Baldwin, Civil Engineer.
Surveyed & Drawn by B.F. Perham.
May 1834

In ceasing to keep his daily journal and replacing this activity with the creating and filling out of the charts of general phenomena, Thoreau was turning away from linear time—of which he knew there was likely little left—and toward the unending cycles of seasonal time. This was the same shift he'd arrived at after the death of his brother, and that he'd made many times since in the wake of other losses, albeit in less comprehensive ways. Now, facing the loss of his mobility—and with it his ability to immerse himself in the more-than-human world and in the ways he'd built his days around for so long—as well as the prospect of death, Thoreau sought another way to live in and through the seasons: by meticulously charting their most distinct phenomena. The technology he'd once embraced for its power to increase his perception and comprehension of the more-than-human world now served a new function: in his final months, the Kalendar offered the confined and disabled Thoreau a new, virtual experience of the seasons.

Filling in the categories for the October chart, Thoreau revisited his many journal entries about the river, mining them for measurements and insights about what one friend called "the river that ran through his life." Under the column for 1852, he writes, "Oct 2nd 'within an inch or 2 as low as when I made my mark' Aug. 23 8 9 1/2 /12ths ft [8 feet, 9½ inches] lower than in April or (as I calc. in 61) 2 inch lower s.l. 59."[24] This complicated notation traces the relationship between several, interrelated observations: he compares the level of the river on October 2nd of 1852 (this date corresponds to the chart location) to that of August 23, 1852, the date on which Thoreau "made [his] mark" on the railroad truss. He also notes April 23, 1852, the high-water mark for the river that year, and the "s.l." (summer level) for 1859. Remarkably, in parentheses, he also records the present moment of his calculations—"as I calc.[ulate] in 61"—and hence of his creation of the chart. Encountering this entry—the cramped writing filling the tiny box with this jumble of dates and measurements—one can feel the appeal of the project for Thoreau at this time in his life: in a single moment, he was able to grasp the whole of the river's flux over time. Unable to venture out onto the water, he could know the river this way, through the permutations of its *temporal* flow.

Intriguingly, the river measurements continue long after the journal entries end: the boxes here are full through November 1861. Given what we know of Thoreau's condition at this time, it seems likely that it was Channing who was making these measurements, and making them often—stopping by the willow tree where he and his friend had launched countless adventures, but would no more, picking up the measuring stick, and noting down the day's level. When he visited Thoreau, Channing came bearing gifts: news of a late-blooming strawberry, the pulse of the living waters.

Dying, Thoreau had to imagine new ways to be in touch with the more-than-human world that was the central source of meaning and solace in his life. These ways involved the construction of a virtual nature through the charts of general phenomena. They also involved a letting go of the models of self-reliance and independence he'd once espoused. Darwin and Brown had taught him that interdependence was the way of things, both in nature and in moral life. The limits of his body now forced him to live this principle, accepting his dependence on others: his sister Sophia, who acted as editor, agent, nurse, and amanuensis in these final months, and his friends, who brought the seasons to him.

The story of Thoreau's practices of being with and representing the more-than-human world—his enlistment of writing, walking, measuring, charting, and mapmaking, all of them bearing the traces of his particular life while seeking ever wider and more comprehensive views—is a helpful one for our own alienated, grief-struck, and damaged age. As our access to data about the changing climate, species loss, and injustice around the world has increased exponentially, exceeding even the wild epistemological dreams of the nineteenth century—our ability to truly *know* our connection to the vast systems of which we are part—has not kept up. And while Thoreau's practices can't be ours—they are the practices of his time, place, and culture—we can learn from the instinct that animated them and that drove their evolution: to locate justice and recovery and, finally, solace, in knowing (in all the senses of this word) the necessary links between us and everything else.

Kristen Case is the author of Keeping Time: Henry David Thoreau's Kalendar *(forthcoming) and three books of poetry, most recently* Daphne *(forthcoming). She lives in Farmington, Maine.*

Notes

Epigraph 1: Quotations for Thoreau's journal are taken from the Princeton University Press edition where possible. They have published eight volumes to date, encompassing entries from October 22, 1837, to September 3, 1854. Later quotations are taken from the 1906 Houghton Mifflin edition edited by Bradford Torrey and Francis Allen. The Princeton edition is cited in the text parenthetically as PJ, followed by volume and page numbers. The Houghton Mifflin edition is cited parenthetically as J, followed by volume and page numbers. Henry David Thoreau, *The Journal of Henry D. Thoreau*, 8 vols. to date (Princeton: Princeton University Press, 1981–). Henry David Thoreau, *The Journal of Henry David Thoreau*, ed. Torrey and Allen, 14 vols. (Boston: Houghton Mifflin, 1906).

Epigraph 2: I am grateful to Rochelle Johnson for drawing my attention to this passage in her haunting and beautiful talk on Thoreau and disability at the Thoreau Society Annual Gathering in July 2022.

1. I periodically use the phrase "more-than-human," coined by David Abram and frequently used in work by posthumanist and ecocritical scholars, to refer to an environment that encompasses both human and nonhuman life. Abram writes, "Caught up in a mass of abstractions, our attention hypnotized by a host of human-made technologies that only reflect us back to ourselves, it is all too easy for us to forget our carnal inherence in a more-than-human matrix of sensations and sensibilities. Our bodies have formed themselves in delicate reciprocity with the manifold textures, sounds, and shapes of an animate earth—our eyes have evolved in subtle interaction with other eyes, as our ears are attuned by their very structure to the howling of wolves and the honking of geese. To shut ourselves off from these other voices, to continue by our lifestyles to condemn these other sensibilities to the oblivion of extinction, is to rob our own senses of their integrity, and to rob our minds of their coherence. We are human only in contact, and conviviality, with what is not human." *The Spell of the Sensuous: Perception and Language in a More-than-Human World* (New York: Vintage, 1997), 22.
2. Henry David Thoreau, *Walden*, ed. J. Lyndon Shanley (1971; Princeton, NJ: Princeton University Press, 2016), 329.
3. Laura Dassow Walls, *The Passage to Cosmos: Alexander von Humboldt and the Shaping of America* (Chicago: University of Chicago Press, 2009), 5.
4. Thoreau, *Walden*, 304.
5. Laura Dassow Walls, *Henry David Thoreau: A Life* (Chicago: University of Chicago Press, 2017), 147.
6. Henry David Thoreau, *The Correspondence of Henry David Thoreau*, ed. Walter Harding and Carl Bode (New York: New York University Press, 1968), 34.
7. Charles Darwin, *On the Origin of Species by Means of Natural Selection; or, The Preservation of Favoured Races in the Struggle for Life* (London: John Murray, 1859), 73.
8. Henry David Thoreau, *Reform Papers*, ed. Wendell Glick (Princeton, NJ: Princeton University Press, 1973), 73.
9. Ibid., 124.
10. Walls, *Thoreau*, 459.
11. Ibid., 458.
12. Darwin, *On the Origin of Species*, 73.
13. Walls, *Thoreau*, 472; Michael Benjamin Berger, *Thoreau's Late Career and "The Dispersion of Seeds": The Saunterer's Synoptic Vision* (Rochester, NY: Camden House, 2000), 4.
14. Henry David Thoreau, charts of general phenomena. Manuscripts are held at the Morgan Library and Museum in New York (April, May, June, November) and the Beinecke Rare Book and Manuscript Library at Yale University (October and December). Transcriptions available at Thoreauskalendar.org.
15. The Clark Museum's exhibition *On the Horizon: Art and Atmosphere in the Nineteenth Century* (November 19, 2022–February 12, 2023), https://www.clarkart.edu/microsites/on-the-horizon/.
16. Henry David Thoreau, *A Week on the Concord and Merrimack Rivers*, ed. William L. Howarth and Elizabeth Hall Witherell (Princeton, NJ: Princeton University Press), 176.
17. Thoreau, *Walden*, 131.
18. Ibid., 123.

19. Henry David Thoreau, Nature notes, charts, and tables: Autograph manuscript, 1851–60. The Morgan Library and Museum, MA 610.
20. Daegan Miller, *This Radical Land: A Natural History of American Dissent* (Chicago: University of Chicago Press, 2018), 23.
21. Robert M. Thorson, *The Boatman: Henry David Thoreau's River Years* (Cambridge, MA: Harvard University Press, 2017), 23.
22. Miller, *This Radical Land*, 39.
23. H. Daniel Peck, *Thoreau's Morning Work: Memory and Perception in "A Week on the Concord and Merrimack Rivers," the Journal, and "Walden"* (New Haven: Yale University Press, 1990), 75.
24. Thoreau, Nature notes (General Phenomena for October).

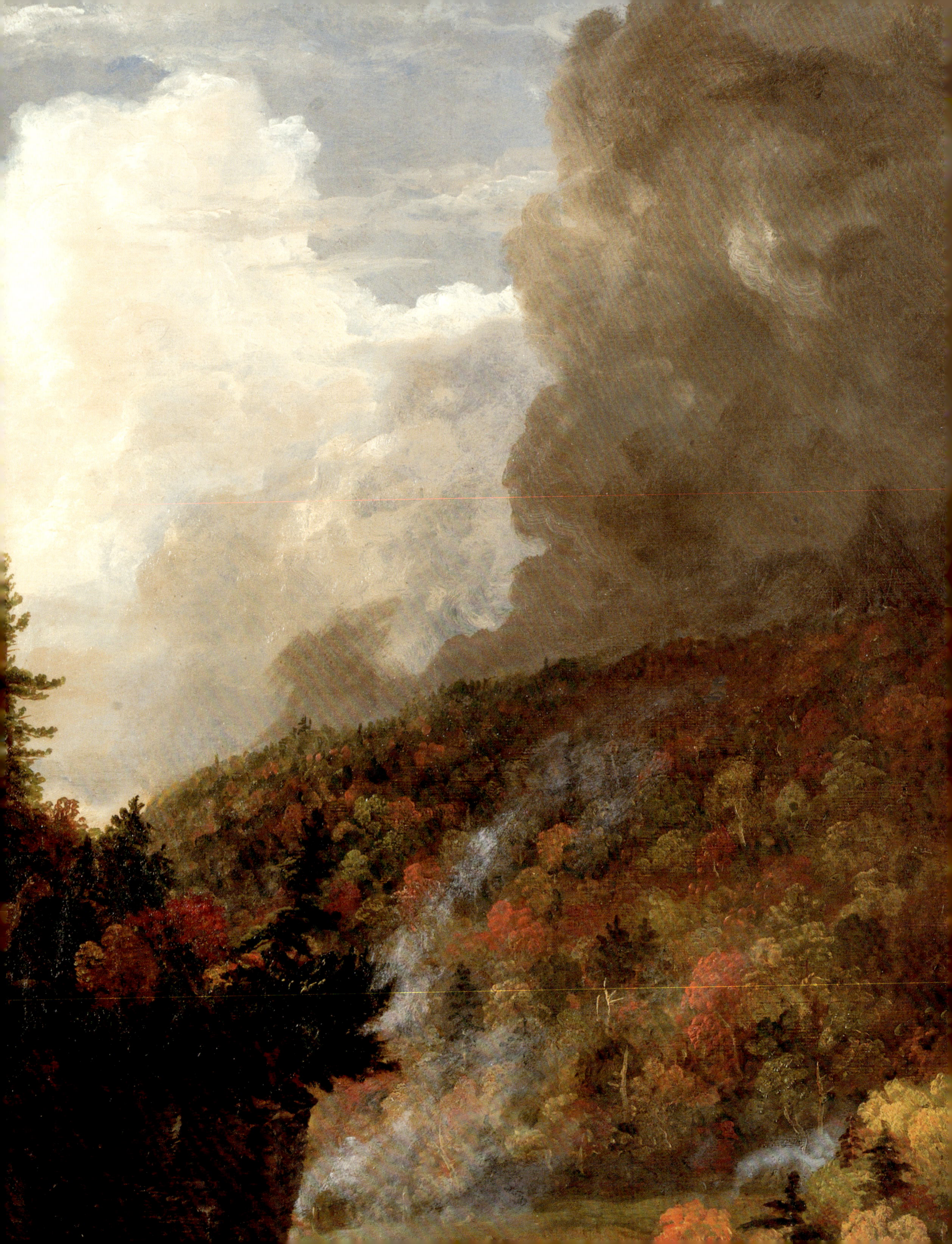

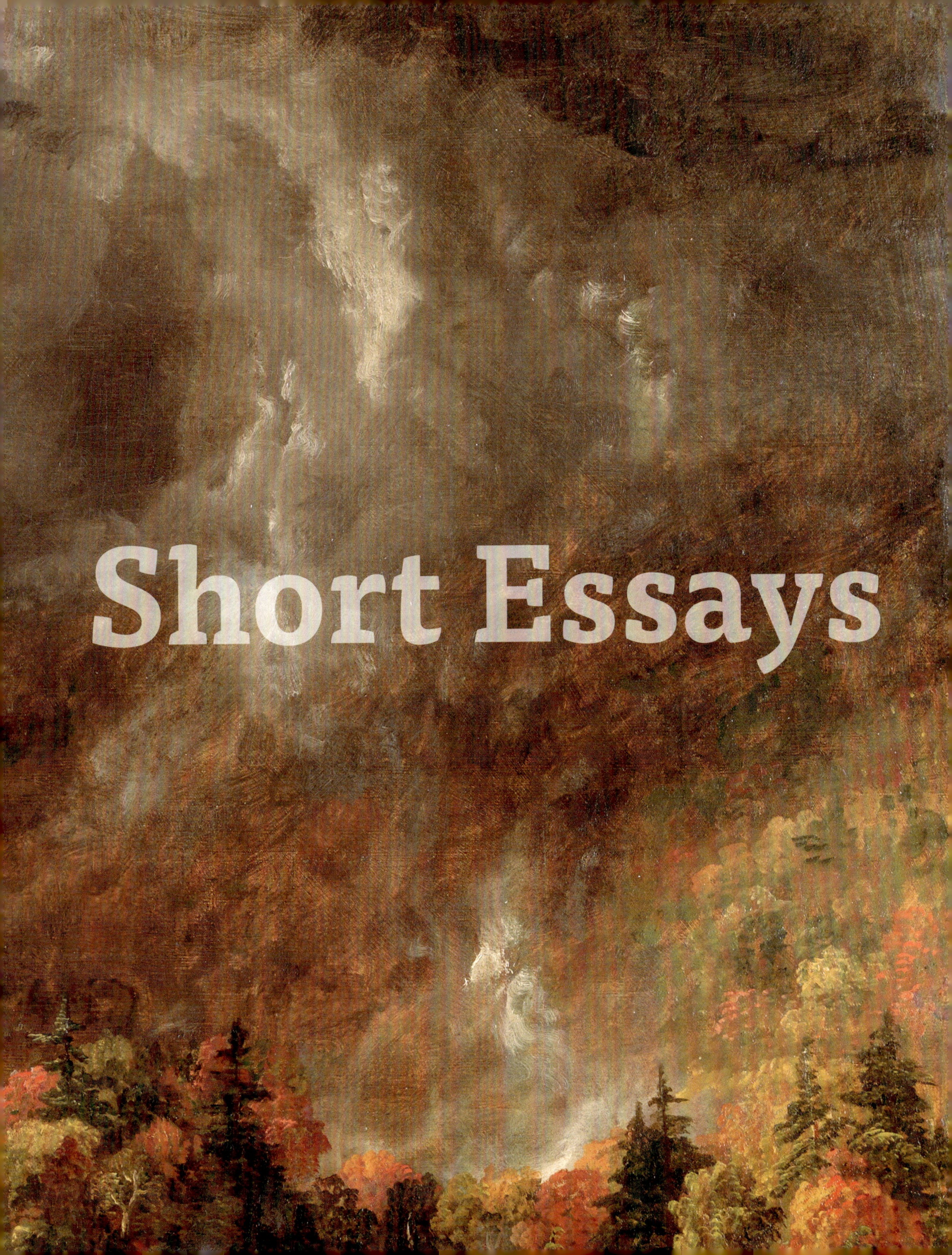
Short Essays

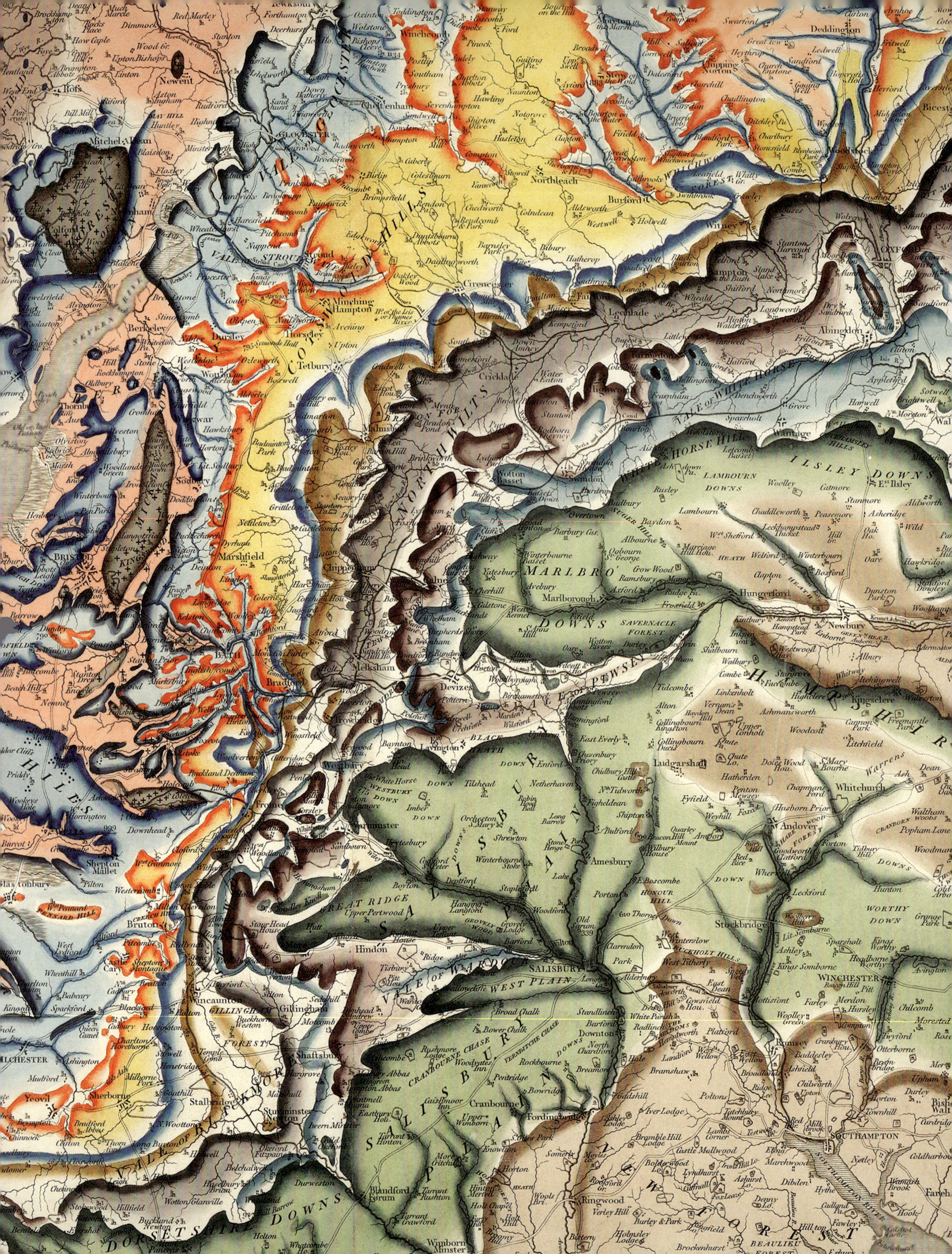

COTSWOLD HILLS
NORTH WILTS
VALE OF WHITE HORSE
WHITE HORSE HILL
ILSLEY DOWNS
LAMBOURN DOWNS
MARLBRO DOWNS
SAVERNACLE FOREST
PEWSEY
SALISBURY PLAIN
GREAT RIDGE
WEST PLAIN
CRANBOURNE CHASE
DORSETSHIRE DOWNS
NEW FOREST
HAMPSHIRE
Deddington
Bicester
Newent
Ross
Cheltenham
Gloucester
Northleach
Burford
Witney
Oxford
Abingdon
Stroud
Cirencester
Dursley
Tetbury
Berkeley
Wantage
Swindon
Marlborough
Hungerford
Newbury
Chippenham
Marshfield
Melksham
Devizes
Kingsclere
Whitchurch
Andover
Amesbury
Warminster
Heytesbury
Frome
Wells
Bruton
Hindon
Salisbury
Stockbridge
Winchester
Romsey
Southampton
Wincanton
Shaftsbury
Sherborne
Yeovil
Blandford
Ringwood

Jan Zalasiewicz

Patterns of Strata

William Smith played a giant role in the history of geology. His work is a mirror to the science itself, in both revealing the deep rock structure underground and, as an avowedly practical man, in helping exploit its resources.

His achievement still seems impossible, absurd: to map the geology of a whole country, single-handedly, from scratch, on foot and by horse-drawn carriage—and all as a largely self-taught working man. He was a surveyor, with a living to earn and a wife to support, in stark contrast with the mostly independently wealthy pioneers of geology. Yet, amid his day-to-day worries (which included a spell in a debtors' prison), his insights and sheer doggedness illuminated this science's foundations.

Anyone who has tried geological mapping knows just how difficult it is to peer beneath a landscape that is almost entirely shrouded by a thick soil layer and clothed in vegetation—as is typical of Britain—and divine the rocky skeleton beneath. I vividly remember my own bafflement as a student first developing this skill: the landscapes I was trying to decipher seemed so obstinately *opaque*. Moreover, Smith had, to a large extent, to work out for himself what became the primary tools of the geologist's trade: the ordering of geological strata glimpsed in the canal and road cuttings he worked on, and the use of fossils as time markers within strata (see image on page 102). He became adept at "reading" landscapes, tracing out ridges formed by hard rock layers and the intervening hollows that betray softer-weathering strata. His vivid, colorful maps and the geological cross sections he drew, conveyed, at a glance, both the surface expression (or "outcrop"), and the three-dimensional architecture of the ground, with its progression of strata made logical and predictable (see image on the facing page and page 103). His insights also helped change the trajectory of Earth history.

Smith was a practical man, and his maps were meant to be useful. The order they created was intended to guide the discovery of coal seams: to predict their locations underground by the logic of the strata, rather than by guesswork and hope. This logic was later extended to exploration for oil, gas, and other natural resources. The study of geology has become enormously sophisticated, using computers and operating at a speed that can allow the courses of individual boreholes to be tracked and modified, even as they are drilled. Yet it remains based on the fundamental principles that William Smith developed.

This practical approach to geology has been strikingly successful, and the resources generated have driven the modern world. In turn, this has created untold wealth, but the consequences include a perturbation of Earth's carbon cycle, which in its scale and speed is without known precedent in our planet's history. Humanity has, perforce, taken the reins as climate driver, and is rapidly shaping a world different from that in which Smith lived. His legacy, thus, is double-edged, with each edge becoming sharper as each year passes.

Jan Zalasiewicz is Emeritus Professor of Palaeobiology at the University of Leicester and a member of the Anthropocene Working Group. His research focuses on fossil ecosystems and environments.

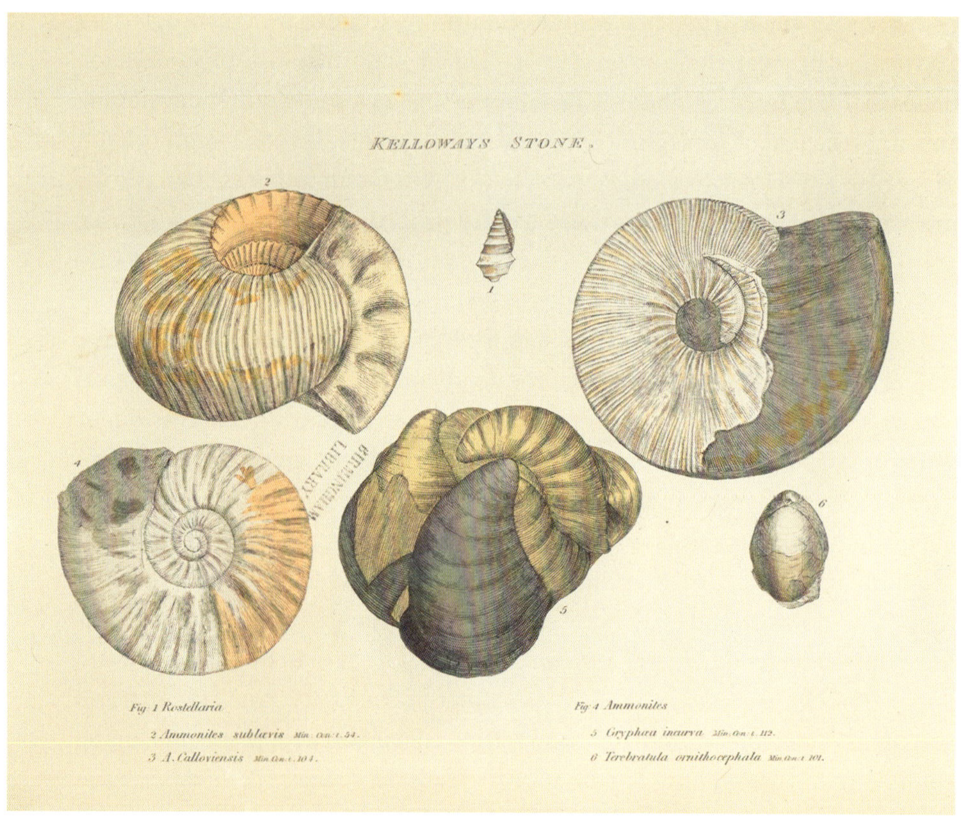

"Kelloways Stone" by James Sowerby in William Smith, *Strata Identified by Organized Fossils* (London: W. Arding, c. 1816–19), plate 14. The Huntington, 474723.

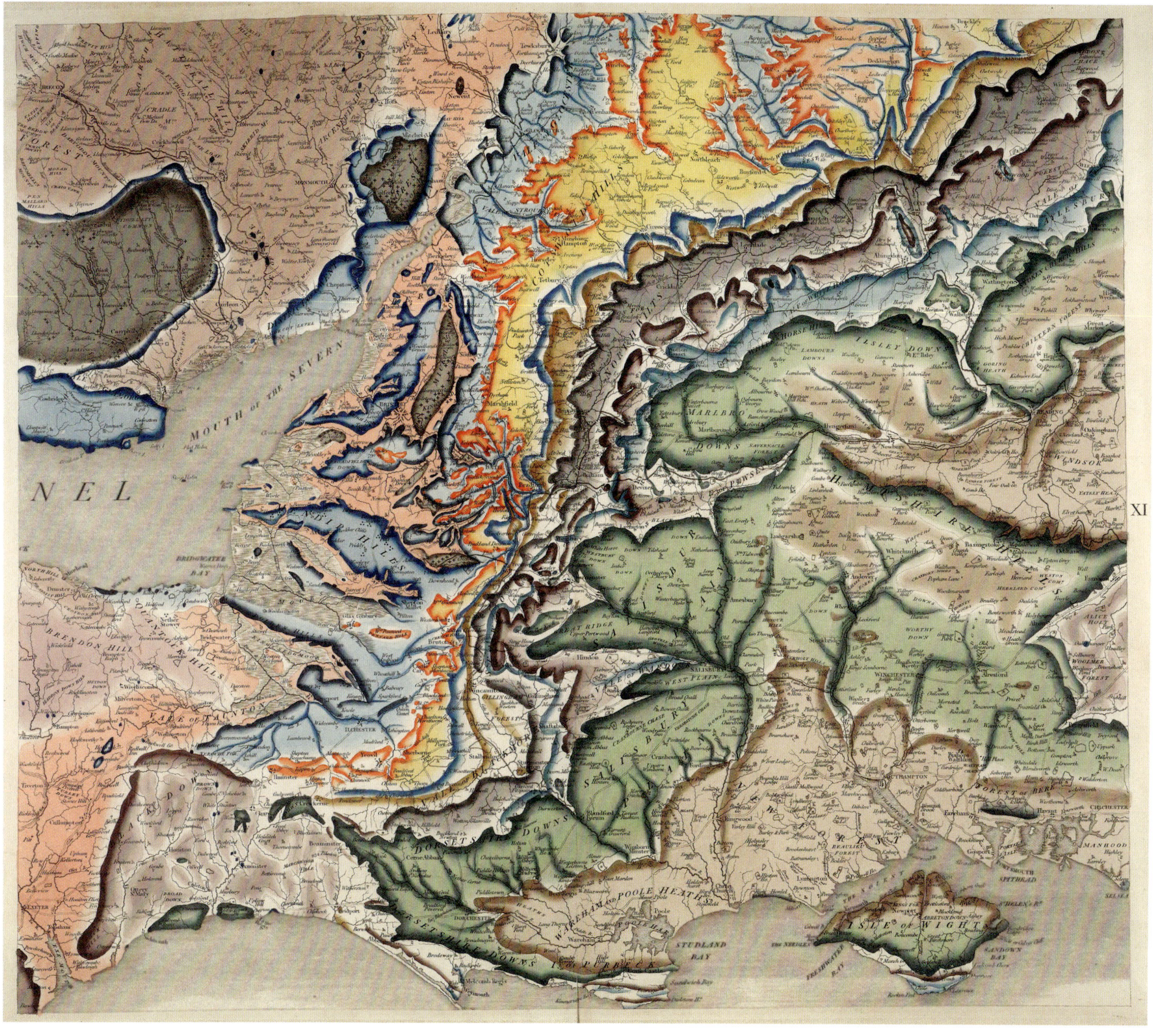

Detail of William Smith, *A Delineation of the Strata of England and Wales, with Part of Scotland* (London: J. Carey, 1815), map XI. The Huntington, 476026.

Working as a surveyor in southwest England, Smith spent years observing coal mines and canal excavations. He noticed that rock layers were arranged in specific patterns always in the same relative position, and realized these strata could be identified and dated by the fossils each contained, with more recent layers nearer the surface. Smith completed a survey of the entire country and published the first geological map of Great Britain in 1815. His work contributed to our understanding of the vast age of the Earth and the continuing extinction and evolution of species over time.

Gillian Osborne

Imagining Distorted Scale

Learning from the Hitchcocks

In his popular textbook, *Elementary Geology,* nineteenth-century American scientist, educator, and minister Edward Hitchcock described how images of geologic strata must be "somewhat ideal." While these rock layers can occasionally be directly observed along "cliffs, on the sea coast, or the banks of rivers"—more often they are buried below the Earth's surface, where only a picture will bring them into sight. But while other objects of natural history—plants or fossils, say—can be rendered close to their true size, the constraint of translating a river system, mountain range, state, or country to paper, results in what Hitchcock called a "distortion" between visual axes; to accommodate Massachusetts or England, the highest hills must be reduced to bumps.[1]

Such a vision of scale—both ideal and out of whack—is evident in *Drawing of Strata across England* (shown below), one of many instructional illustrations produced by scientist, educator, painter, and wife of Edward Hitchcock, Orra White Hitchcock. "It is not in my power to describe these scenes with the skill of the poet or the painter," Edward wrote in one of his earliest studies of the Connecticut River Valley, which would become the site of his most consequential discovery—the tracks of what he believed to be ancient birds.[2] He relied on Orra to illustrate his lectures and essays as professor and eventually president of Amherst College.

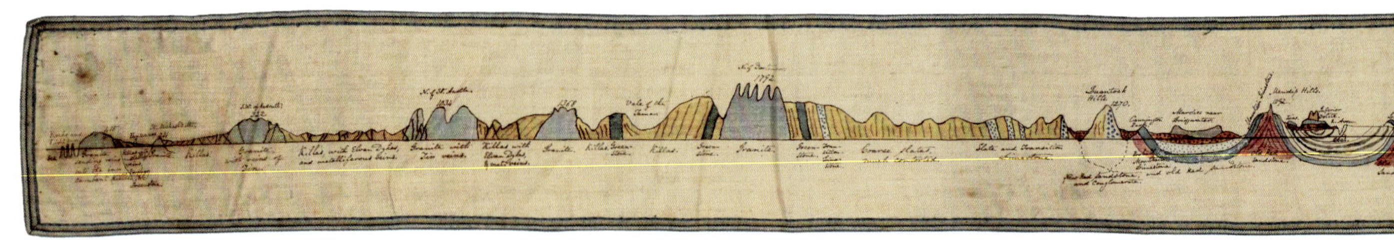

Edward found Orra's classroom posters to be "indispensable" learning "aids." These, along with "cabinets," collections of fossils and specimens, gave students "knowledge of the natural productions of different parts of the earth." Image, combined with object, expanded their imaginations from the local to the global, and helped them understand themselves in relation to the long history of the Earth.[3]

Hitchcock joined other nineteenth-century natural theologians in applying recent geologic discoveries to new interpretations of biblical and human history.[4] And yet Hitchcock was distinctive in his sense of humans as direct actors in geologic history—both on a material, and a spiritual, plane. At the end of his career, Hitchcock described the universe as a reverberating "telegraphic system," in which "every atom" is "a living witness to the actions of every living being."[5] The weirdness of this—and the need for faculties beyond direct observation to intuit such echoes—should not be overlooked. Like strata buried below the Earth's crust, telegrams from the universe need to be imaginatively received.

Orra White Hitchcock, *Drawing of Strata across England*, c. 1828–40. Pen and ink on linen, 5⅛ × 63 in. (13 × 160 cm). Courtesy of Amherst College Archives and Special Collections, Orra White Hitchcock Classroom Drawings Collection, Map Case 3, Drawer 14.

Used as classroom aides for her husband Edward's lectures, Orra White Hitchcock's drawings helped geology students at Amherst College visualize scientific concepts. Painted on poster-size pieces of cloth, with bright colors and bold outlines, they were meant to be seen and easily comprehended from a distance. This long, thin illustration depicts the strata of England's south, viewed in cross section from the west to east coasts. Images like this attest to Orra's engagement with new scientific developments, such as William Smith's efforts to map the geology of England, an understanding of which she effectively communicates here.

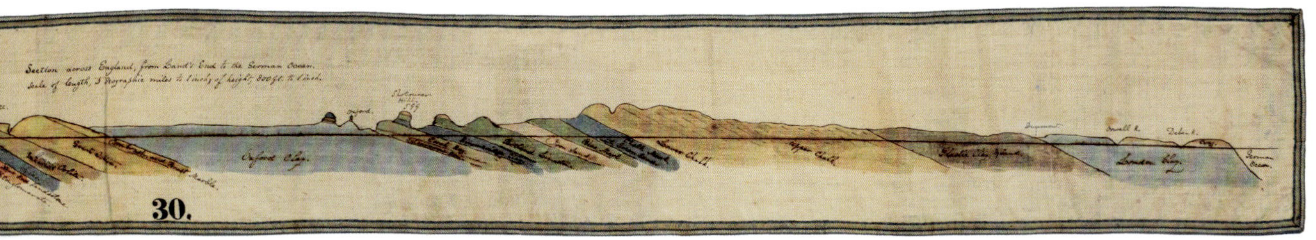

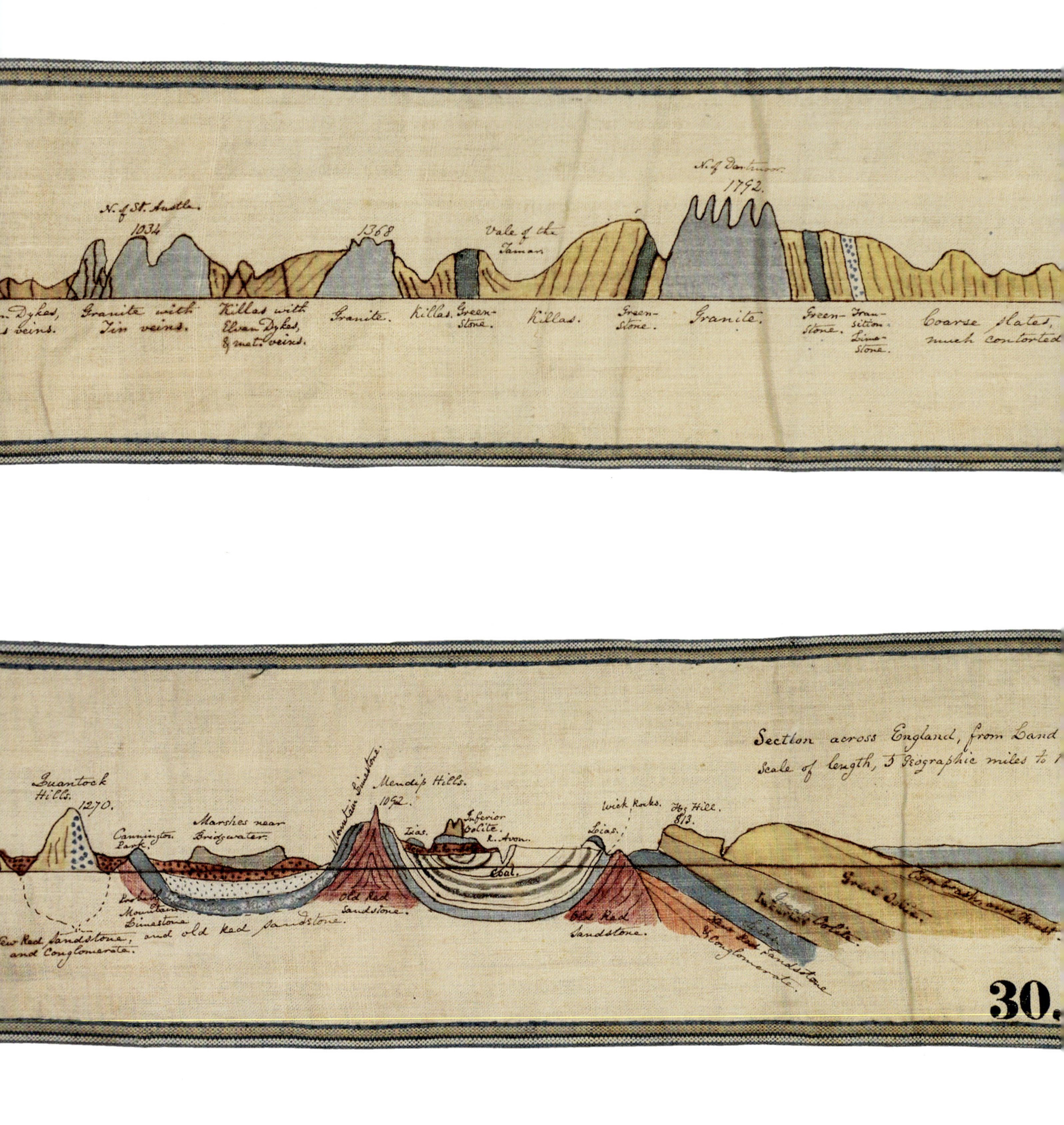
N. of St. Austle
1034
1368
Vale of the Tamar
N. of Dartmoor
1792.
Granite with Tin veins.
Killas with Elvan Dykes, & met. veins.
Granite.
Killas.
Green-stone.
Killas.
Green-stone.
Granite.
Green-stone.
Transition Lime-stone.
Coarse slates, much contorted
Section across England, from Land
Quantock Hills. 1270.
Cannington Park.
Marshes near Bridgwater.
Mountain Limestone.
Mendip Hills. 1092.
Lias.
Inferior Oolite.
R. Avon.
Coal.
Old Red Sandstone.
Mountain Limestone, and old Red Sandstone.
and Conglomerate.
Wick Rocks.
Lias.
Hog Hill. 813.
Old Red Sandstone.
Lias.
Inferior Oolite.
Great Oolite.
& Conglomerate.
30.

On the one hand, Hitchcock's views of human geologic agency anticipate present ideas of the Anthropocene, and the work of his nearer contemporary, George Perkins Marsh. "The human race produce geological changes," Hitchcock wrote in the 1840s.[6] Two decades later, Marsh argued, "it is certain that man has done much to mould the form of the earth's surface," and, "man is everywhere a disturbing agent."[7] On the other hand, where Marsh emphasized the destructive powers of human industry, Hitchcock maintained that humans were artists *as well as* destroyers. Even as humans have eradicated "vast numbers of animals and plants to make room for themselves," "resist[ed] the encroachments of rivers and the ocean," and even affected "climate," Hitchcock wrote, they are also defined in the geologic record by the "peculiar fossil relics" of "the various productions of [their] art."[8]

Hitchcock's sense of the necessity of art, and the necessary distortions of Orra's images, can still teach us today about how to conceive the wonky vectors that connect present to past, or planet to place. The lesson of the Hitchcocks' collaborations is that an environmental imagination can be oblique, and adorned. And that we must draw on all our powers of observation and artistry—as wild as the line between them might be—to see the Earth for what it really is, was, or might become.

Gillian Osborne is a writer and educator. Her book of essays Green Green Green *(2021) draws on archival research on Edward and Orra White Hitchcock.*

M Jackson

Hoping for John Brett's *Glacier of Rosenlaui*

I haven't seen a glacier like John Brett's in my lifetime. I'm forty. I've been a glaciologist my whole career. My son, Finnegan, he's one. And I know he won't ever see a glacier like Brett's in his life.

Some days I debate if I should even teach Finnegan the word, "glacier." I wonder when he'll be ready to conceptualize the scale of such loss.

I don't know.

So instead, I pull Finnegan deeper into my lap and show him my image of *Glacier of Rosenlaui* (on the facing page).

I show Finnegan how in Brett's painting this glacier is wild and alive. Vibrant. She's thick at her edge—what we call the "terminus"—not thinning and brittle, like the termini of most of our two hundred thousand glaciers remaining worldwide today. Brett's glacier is growing, flowing forward over the foliated gneiss ubiquitous in the Swiss Alps. There is no telltale apron of raw earth about her, those planetary auras of exposed crust, evidence of ice dissolving inward far too quickly. Instead, the ice near her terminus is bright and white with hints of blue—free of dark shadows from sediments recently uncovered and blown back over the glacier's body. Free of splashes of neon from wind-carried microplastics. Free of acres of pitch-black smears, aeolian aerosols from far-off coal-fired power plants.

Farther up Brett's glacier I'm heartened to see thick blankets of pale white snow in the accumulation zone. I whisper "accumulation" to Finnegan. He replies, "gah."

The accumulation zone is the glacier's refrigerator, where snow stacks up from winter storms and persists, transforming into glacier ice over years, decades, centuries. This zone is the first place I look on any glacier I work with in North America, Africa, Europe, Middle East, Antarctica. I assess: is there enough snow to feed this glacier over time? Today, almost always, the answer is no. Our glaciers are starving, emaciated skeletons dissolving away before us.

But not Brett's.

Finnegan and I marvel at this glacier's health. My eyes linger on those boulders in front—termed in glaciology "erratics"—and how they have sharp edges. It means they rode the surface of the glacier like a conveyer belt, were dumped

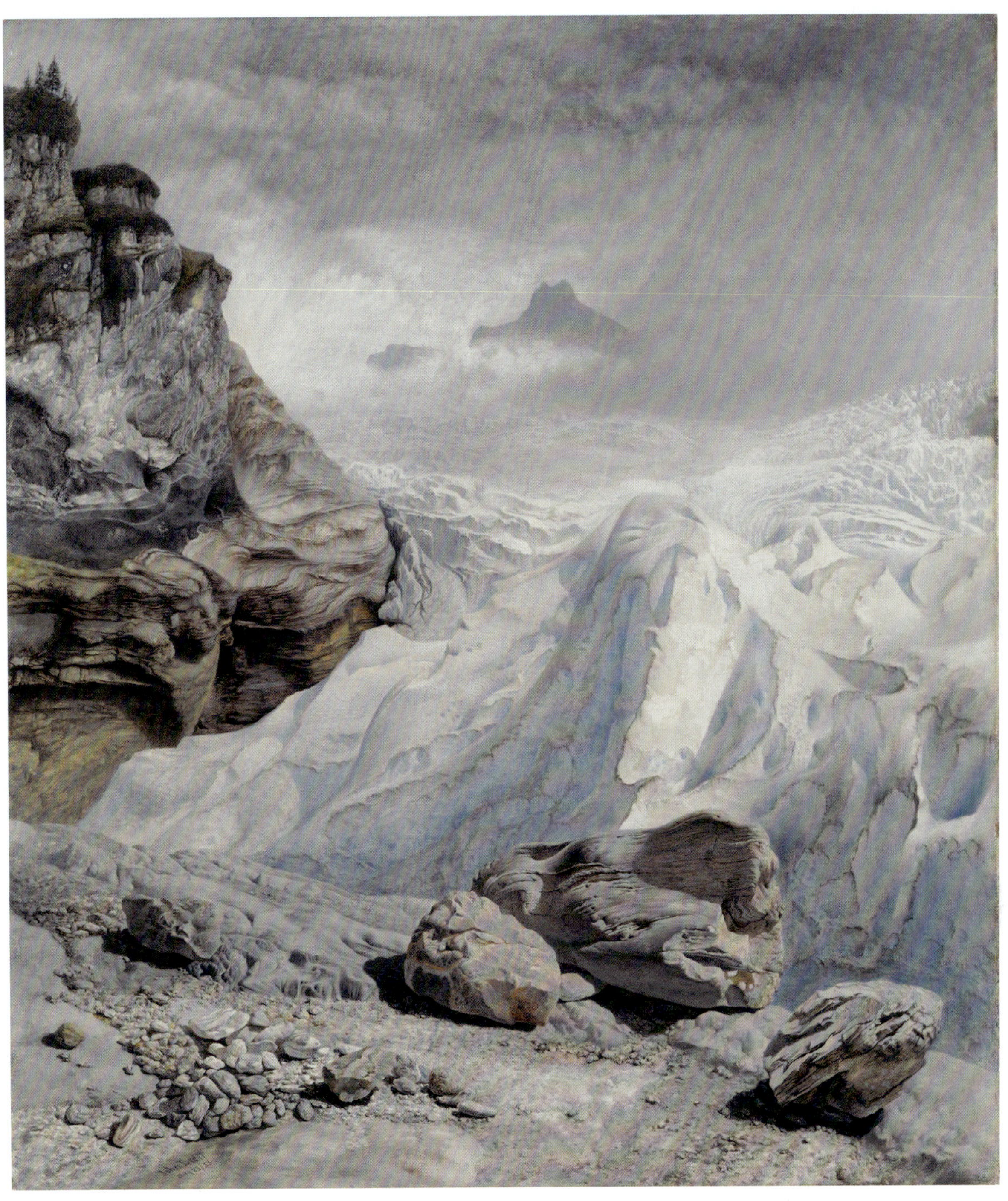

John Brett, *Glacier of Rosenlaui,* 1856. Oil on canvas, 17½ × 16½ in. (44.5 × 41.9 cm). Tate, London, purchased in 1946, N05643.

Inspired by new discoveries in alpine geology, Brett's painting of a Swiss glacier is so precisely rendered one can identify the loose stones littering its path as gneiss, a type of metamorphic rock. The painting effectively suggests the scale of the mountainous landscape with a stand of miniscule trees, which in reality are tall conifers, at top left. The path of the glacier has left the earth devoid of growth, a sign of powerful natural forces at work.

out front like lost airport luggage. Compare them to the smaller rocks in the foreground. Most of those are rounded, like they've been gently tumbled over and over inside, or under, the ice. Or were moved along by that quiet stream flowing near them. Today, those quiet streams are torrents, thick slurries of ice turned to water, full of rounded boulders and glacial till and petrified forests. You hear the roaring from miles away.

I point out to Finnegan the trees in the upper left of Brett's painting. I squint, try to make out the species of tree. Are those cones pointing up for firs, or down for spruces? Brett's brushstrokes are incredibly fine. I guess "spruce." Finnegan nods sagely, repeats, "gah."

What I notice though is that Brett's trees are standing tall. Not bent. As our atmosphere unnaturally heats up, hot air collides with cold air surrounding ice and catalyzes katabatic winds. These are hotter and more powerful than any we've previously experienced, capable of battering the strongest tree and bringing even the most resilient alpine forest to its knees.

Finnegan coos at the glacier, and I smile. Brett's glacier is the one I want my son to know, imagine, remember. So later, after snack, and diaper, and cuddle, I put Finnegan down to sleep. And above his crib, I tape my image of John Brett's *Glacier of Rosenlaui*, so he'll always know what our future could look like, one day, with healthy glaciers again.

M Jackson is a glaciologist and author. Her books The Secret Lives of Glaciers *(2019) and* The Ice Sings Back *(2023) explore the societal impacts of climate change.*

Karen Lloyd

Viewing Stations in the English Lake District

First published in 1778, Thomas West's *A Guide to the Lakes in Cumberland, Westmoreland, and Lancashire* (on page 113) provided a resource for tourists who decided to travel within England rather than undertake a grand tour of Continental Europe. Establishing a series of "viewing stations," West's guide functioned as a kind of eighteenth-century Tripadvisor, pointing visitors to the most sublime views in the region, such as those in the vales of Rydal and Grasmere. He also instructs visitors exactly how to look. At each viewing station West invites them to turn their backs on the landscape itself and look instead through a Claude glass, or pocket "landscape mirror" carried specifically for this purpose. Turning the view into a reflected picture in this way, West suggests, will enhance the viewer's perspective and experience.

Painted almost two decades after the publication of West's guide in 1778, J. M. W. Turner's watercolor *Grasmere, Looking South towards Rydal Water* (on page 112) invites us to see a landscape almost devoid of human presence that, in reality, would have been a place of pastoral industry, thus circumscribing the view much like West's Claude glass. Rydal and Grasmere are synonymous with the Romantic movement and with William Wordsworth in particular, whose poetry emphasizes not only the importance of intimate and direct engagement with the local environment but also the "increasingly disruptive influence of man" on that environment.[1] It is useful, therefore, to consider the various ways the Lakes are seen from the metaphorical "viewing station" of the Anthropocene.

By designating the Lake District as a "cultural landscape," the 2017 World Heritage inscription stamped a preservation order on the agro-pastoral tradition of sheep farming practiced in the region for centuries, whereby sheep (such as the Herdwick) are hefted to common upland grazings. But in the postwar period the density of sheep stocks and the resulting overgrazing led to a catastrophic loss of wildlife and habitats, leaving the uplands mostly devoid of the species that should be there—from golden eagles to wildflowers and pollinators. During Storm Desmond in 2015, fifteen trillion liters of rain fell on the lakes in twenty-four hours, destroying more than one hundred bridges and roads, leaving communities divided—in some cases, for more than five years. While tensions exist between the more traditional farming community and the aspirations of conservationists in the Lake District, when faced with the twin effects of biodiversity loss and extreme weather, it feels increasingly urgent to view the land and its ecosystems less in terms of preservation and more through the exigencies of the climate emergency.

When sheep stocks are significantly reduced, the land develops greater complexity. Soils absorb and slow the flow of rain and simultaneously give rise to healthier ecosystems and habitats of meadows, trees, and scrubs, which in turn allow the return of a cascade of species from wildflowers to pollinators, birds, and mammals. Despite the preservation ethic, change *is* underway. An increasing number of farms have transitioned away from sheep to practice regenerative farming methods, including rotating grazing by cattle and resting the ground. Landowners such as those at Lowther near Penrith are undertaking restoration projects across their entire estates from valleys to mountaintops, and projects such as Restoring Hardknott Forest are removing conifers planted as a crop in the uplands, replacing them with diverse native woodlands. With changes such as these underway, we humans acknowledge that we have turned back to face the landscape itself and respond to what it is so clearly showing us.

Karen Lloyd is an author, poet, and environmental activist from the United Kingdom's Lake District and writer-in-residence at Lancaster University's Future Places Centre. Her books include Abundance: Nature in Recovery *(2021).*

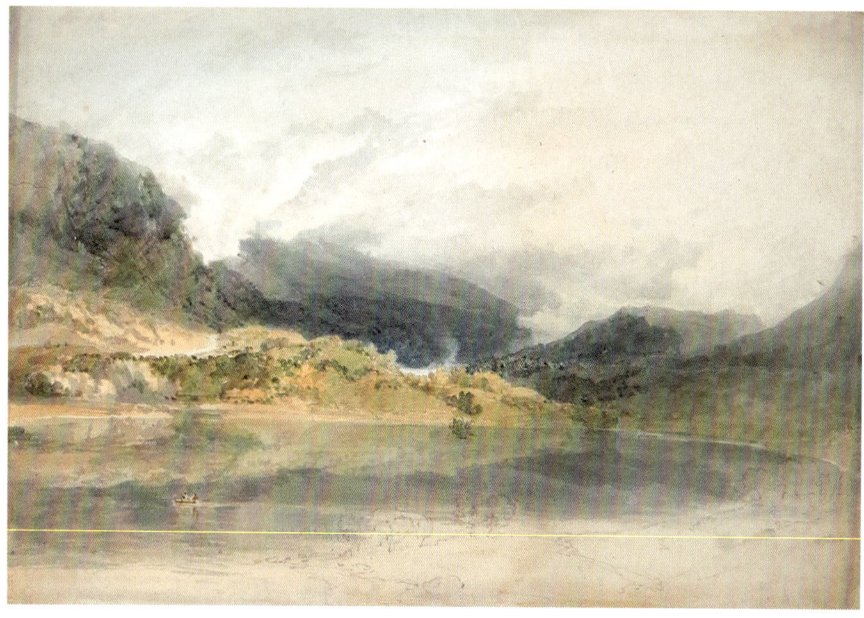

Joseph Mallord William Turner, *Grasmere, Looking South towards Rydal Water,* 1797. Graphite and watercolor on paper, 10¾ × 14½ in. (27.4 × 37 cm). Tate, London, accepted by the nation as part of the Turner Bequest 1856, D01089 Turner Bequest XXXV 87.

80 A GUIDE TO THE LAKES.

the south, at the entrance of the vale, over a noble fore-ground), and commands a charming view of Windermere-water*. The river Rothay winds through the vale, amidst lofty rocks and hanging woods, to join the lake. The road serpentizes upwards, round a bulging rock, fringed with trees, and brings you soon in sight of

RYDAL-WATER,

A lake about a mile in length, spotted with little isles, and which communicates, by a narrow channel, with

GRASMERE-WATER,

The river Rothay is their common outlet.

Mount Grasmere hill, and from the top, have a view of as sweet a scene as travelled eye ever beheld†. Mr. Gray's description of this peaceful, happy vale, will raise a wish in every reader to see so primæval a place.

"The bosom of the mountains, spreading

* The style of this landscape will be seen in No. 15, of Mr. Farrington's views.

† A little to the left of the road, is No. 5, of Mr. Farrington's views.

A GUIDE TO THE LAKES. 81

here into a broad bason, discover in the midst Grasmere water; its margin is hollowed into small bays, with eminences; some of rock, some of soft turf, that half conceal and vary the figure of the little lake they command: from the shore, a low promontory pushes itself far into the water, and on it stands a white village, with a parish church rising in the midst of it; hanging inclosures, corn fields, and meadows, green as an emerald, with their trees, and hedges, and cattle, fill up the whole space from the edge of the water; and just opposite to you, is a large farm house, at the bottom of a steep smooth lawn, embosomed in old woods, which climb half way up the mountains' sides, and discover above, a broken line of crags, that crown the scene. Not a single red tile, no staring gentleman's house, break in upon the repose of this little unsuspected paradise; but all is peace, rusticity and happy poverty, in its neatest, most becoming attire*."

Mr. Gray's description is taken from the road descending from Dunmail-raise. But the more advantageous station, to view this romantic vale from, is on the south end of the western side. Proceed from Ambleside

* The whole of Mr. Gray's journal is given in the Addenda, Article III.

G

"Rydal-Water and Grasmere-Water" in Thomas West, *A Guide to the Lakes in Cumberland, Westmoreland, and Lancashire* (1778; London: W. Pennington, 1821), 80–81. The Huntington, 298519.

First published in 1778, West's guide described the scenery of England's Lake District in a way that would be useful for artists and tourists. The book presents this mountainous region dotted with lakes through a series of viewpoints, called "stations." By encouraging travelers to appreciate the landscape's aesthetic qualities, West was among the first to consider nature as something to be experienced for pleasure. This idea informed later conservationists who sought to preserve areas of natural beauty. Today, the region is one of the United Kingdom's most popular tourist destinations, its landscape and culture protected as a UNESCO World Heritage Site.

Rachel Storer

Skying

Two renditions of cumulus clouds appear from different perspectives—one aiming more for the scientific, one for the artistic—but both embracing beauty and seeking understanding. I've dedicated my career to learning more about cumulus clouds, using computer models and measurements from satellites to try to discover how quickly air moves up in convective clouds, how this changes depending on the environment, and how it affects the weather that results. No matter how long I work on the scientific questions surrounding cumulus clouds though, nothing can take the place of just watching the sky as the clouds grow and change. I believe an appreciation for both their aesthetic value and their scientific complexity are essential. Each enhances the other.

In the study by Luke Howard (shown below), we see what we would now call a cumulonimbus cloud shown with its anvil. Cumulonimbus clouds tower from near the ground to the top of the troposphere, and when they cannot grow vertically anymore, the mass of air and ice spreads out horizontally, forming the anvil shape shown here. These details were only beginning to be understood at the time, but Howard demonstrates them well.

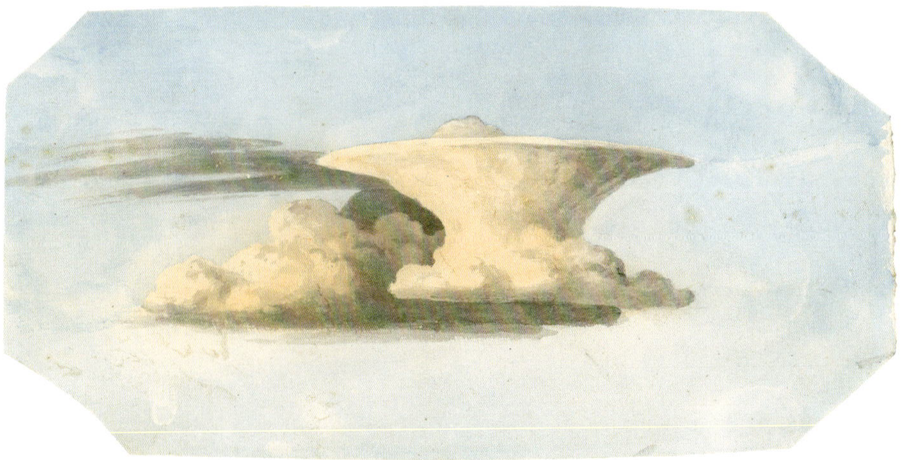

Luke Howard, *Cloud Study of Nimbus Showing Anvil with Cumulus and Water Vapour Streaming Out*, c. 1803–11. Blue, gray, and buff wash with white on paper, overall: 5 × 10 in. (12.5 × 25.4 cm). Royal Meteorological Society / Science Museum Group, 1981–86 /28.

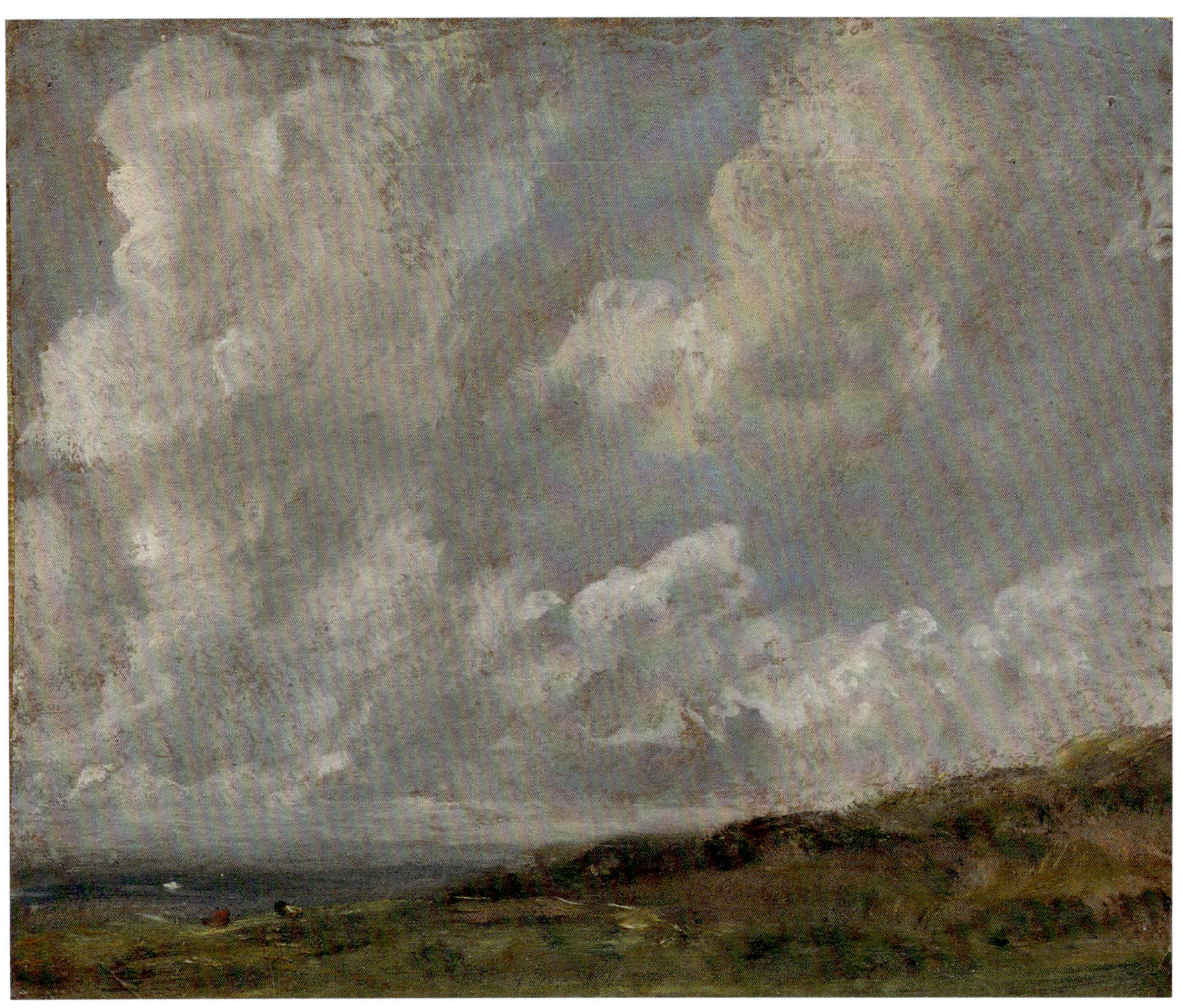

John Constable, *Study of Clouds over a Landscape*, c. 1821–22. Oil on laminate cardboard, mounted on canvas, 9⅝ × 11⅝ in. (24.4 × 29.5 cm). Courtesy of Clark Art Institute, Williamstown, Gift of the Manton Art Foundation in memory of Sir Edwin and Lady Manton, 2007.8.35.

Constable's cloud studies reflect his attempt to align his art with a meteorological understanding of atmospheric phenomena then only recently described by scientists such as Luke Howard. Though not as purely descriptive as Howard's scientific renderings, Constable's cloud studies reveal an understanding of the form and movement of clouds. "Skying," the artist's term for his practice of cloud watching, helped him develop a language of expression for his larger landscapes, where the representation of the sky conveys not only the scene's weather conditions but also the painting's mood.

In the sketch by John Constable (on page 115), the sky is filled with growing cumulus clouds, though they have not quite developed to the stage seen in the Howard study. "I have done a great deal of skying," wrote the artist, referring to the detailed studies of clouds that he made as references for his realistic paintings of the countryside.[1] His realism is stunning. Looking at Constable's oil sketch—at the bubbling, darkening clouds—I can almost feel the humidity of a summer afternoon, the wind picking up with the promise of an evening storm. Constable was influenced by the works of Howard and other contemporary natural scientists who were just beginning to define the different types of clouds, give them names, and explain how they developed. Meteorology was an exciting emerging field of study at the time; early meteorologists pursued versions of the questions I answer with computer modeling with paper and pencil.

I think, or at least I hope, that all of us have done our own version of "skying" at some point. If not, I recommend taking the time to pay attention to the clouds and what they are telling us. Can you see the winds blowing the clouds in different directions or speeds at different heights? Can you see cumulus clouds bubbling up at the top as their buoyancy helps them grow? Once you know what you're looking for, you may also be able to identify the contrails of planes dissipating with time as the air gets mixed and the ice sublimates into vapor, or the virga falling from a cloud on a dry day, never reaching the surface as rain.

The sky also reflects changes brought about by humanity. Those beautiful sunsets, more pink and orange than normal during (increasingly longer) fire seasons, result when smoke scatters away more of the blue light. Maybe all one sees in their skying are puffy clouds shaped like a bunny or a dragon—and that's great, too. What would art be—what would the world be—without some whimsy?

Rachel Storer is a research scientist with UCLA JIFRESSE (Joint Institute for Regional Earth System Science and Engineering) at NASA's Jet Propulsion Laboratory. She studies interactions between deep convective clouds and the environment.

Veront M. Satchell

The Hope Lands, Saint Andrew Parish, Jamaica, in 1826

An Agent of Climate Change

The 1826 map of the Jamaican properties of the Duke and Duchess of Buckingham and Chandos (on page 119) includes the Hope sugar estate, the Merrymans coffee plantation in the Blue Mountains, and the estate wharves in Kingston Harbour. This document represents one of the hundreds of sugar estates that covered the island beginning in the eighteenth century, businesses that made the small British colony and its planters, 85 percent of whom were absentee, the wealthiest in the empire.[1] The system of land use and economic prowess this map represents had serious environmental repercussions that still affect the island today.

When the English captured Jamaica from the Spanish in 1655, apart from a few former Taíno villages and Spanish settlements, it was a dense forest. The invading officers seized large tracts of land as war booty. Initially, large-scale agriculture was not attractive; clearing the forest was daunting and laborers were few. However, the rising demand for sugar in Europe during the late 1600s motivated some early landholders to clear sections of their properties for sugarcane and establish rudimentary sugar mills. These included Major Richard Hope, the original owner of the Hope Estate. To be profitable, sugar production demanded extensive acreages of level land, a large labor force, and herds of cattle to power mills and pull transport carts. Land had to be cleared of vegetation by chopping and burning to make room for planting, pasturage, provision grounds, and residential and factory sites. Trees had to be felled for the timber needed for the various estate buildings, furniture, and wood used as fuel for cooking and boiling cane juice. These early planters initiated what would become wholesale deforestation.

With the growing demand for, and profit to be made from, sugar during the 1700s, the industry grew. Between 1763 and 1786, the number of estates nearly doubled: from 566 to 1,061. With the introduction of coffee, hundreds of large coffee plantations, like the Merrymans plantation (at the upper right), dotted the interior hills and the Blue Mountains, the major watersheds of the island. Jamaica was now the wealthiest colony in the British Empire, accruing annually between £1.5 and £1.8 million (nearly half a billion GBP, or 620 million USD, in today's money) in profit to Britain.[2] All this was achieved through the exploitation of the land and the labor of more than one million Black people imported from Africa by British traders between 1700 and 1807.

The wanton destruction of Jamaica's forests to facilitate the plantation economy had environmental consequences. Average annual temperatures on the island have increased from 24.5°C (76°F) in 1901 to 26.5°C (80°F) in 2019.[3] Droughts in Jamaica are more intense and prolonged, causing serious harm to its agrarian economy.[4] Storms are increasingly more violent and destructive, such as Hurricane Gilbert, a category five hurricane that hit the island directly in 1988. The resulting damages—to agriculture, infrastructure, housing, and the like—totaled $800 million.[5] The storm clouds engendered by the economic systems represented by this 1826 map continue to rain down torrents of environmental challenges upon Jamaica.

Veront Satchell is professor of economic history at the University of the West Indies, Mona. He is the author of Hope Transformed: A Historical Sketch of the Hope Landscape, St Andrew, Jamaica, 1660–1960 *(2012).*

Edward McGeachy, *New and General Plan of the Hope Estate in the Parishes of Kingston and St. Andrew, Jamaica*, 1826. Unfolded: 47½ × 30 in. (120. 7 × 76.2 cm). The Huntington, Stowe papers, 1175–1919 (bulk 1600–1900), mssST West Indies.

Established by Major Richard Hope in the 1660s when the English army arrived in Jamaica, the Hope Estate grew into a successful producer of sugar, rum, and molasses, requiring the forced labor of nearly four hundred enslaved people. The number had risen to 574 by 1832, the year before slavery was abolished throughout the British colonies. This map shows the estate in 1826. It includes the aqueduct (labeled "gutter") that diverted water from the river to the estate's water-powered sugar works. Portions of this structure remain on the grounds of the Hope Botanical Gardens and the University of the West Indies, Mona.

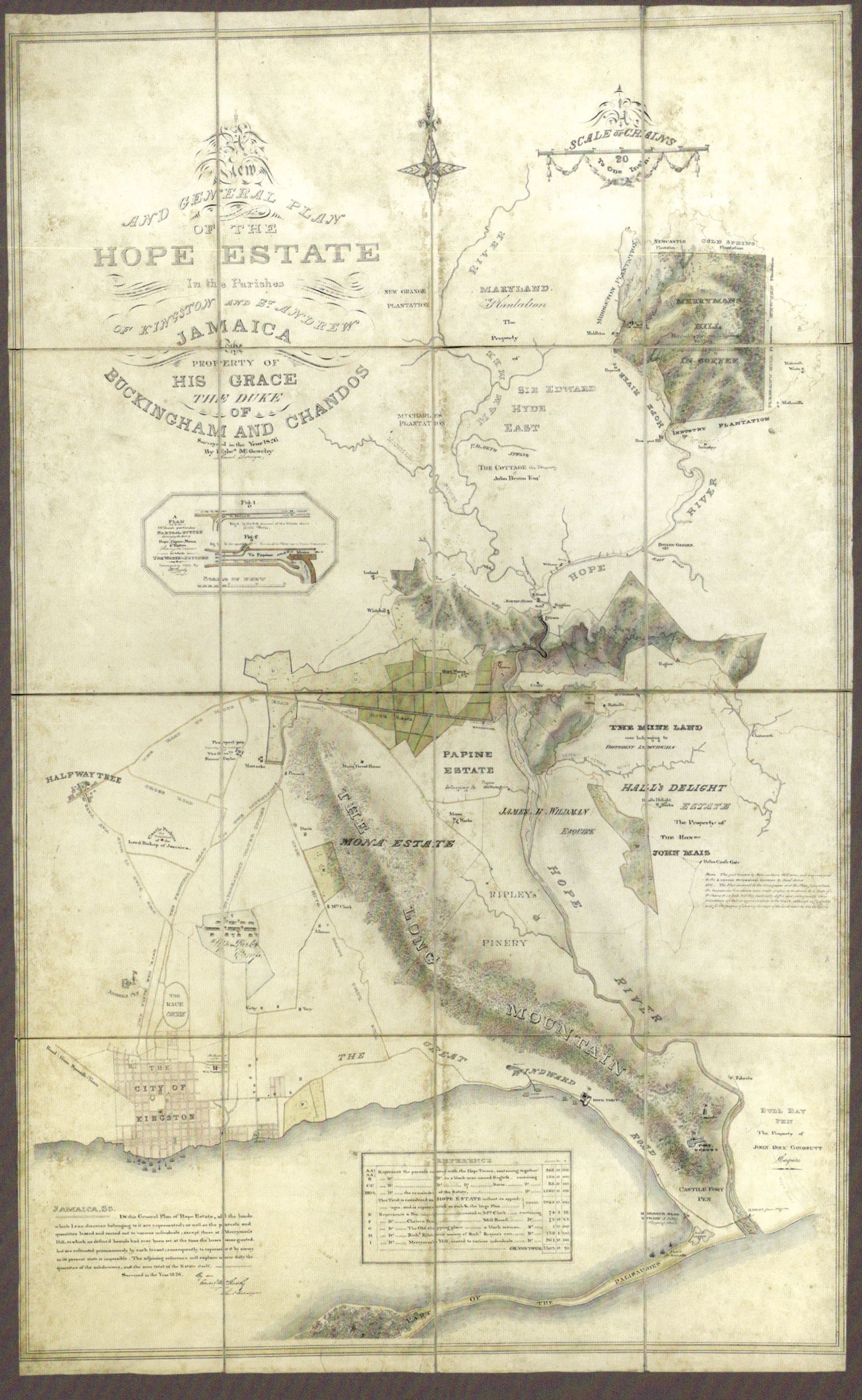

New
AND GENERAL PLAN
OF THE
HOPE ESTATE
In the Parishes
OF KINGSTON AND St. ANDREW
JAMAICA
The
PROPERTY OF
HIS GRACE
THE DUKE
OF
BUCKINGHAM AND CHANDOS
Surveyed in the Year 1826
By Edwd. McGeachy
SCALE OF CHAINS
20 To One Inch
NEW GRANGE PLANTATION
MARYLAND Plantation
RIVER
SIR EDWARD HYDE EAST
MERRYMANS HILL
INDUSTRY PLANTATION
HOPE RIVER
THE MINE LAND
PAPINE ESTATE
HALL'S DELIGHT ESTATE
JAMES W. WILDMAN ESQUIRE
JOHN MAIS
HALFWAY TREE
THE MONA ESTATE
RIPLEY'S
PINERY
LONG MOUNTAIN
THE RACE COURSE
THE CITY OF KINGSTON
THE GREAT WINDWARD ROAD
CASTLE FORT PEN
REFERENCES
JAMAICA, SS.

Suzanne Pierre

There Is No Painting Over a Colonized Ecosystem

When viewing Frederic Edwin Church's *Vale of St. Thomas, Jamaica* (on the facing page), I experience a sense of unease. Across a vast and undulating landscape, dense with broad and deep-green foliage, sunlight casts golden rays that signal either daybreak or the golden hour before twilight. Because I cannot be sure what time of day the painting represents, my attention goes to the topography of the landscape, which centers on a placid river running through the middle of a wide valley. The colors associated with waning and waxing sunlight reveal the montane features. Forests blanketing the mountains and the valley appear to be teeming with life. Despite the equatorial geography of Jamaica, trees resembling the varieties common in forests of the global north dot the mountainsides and the riverbanks. I notice that, from the viewer's godlike vantage point, the Edenic river valley appears in its entirety while the viewer maintains sight of the numerous valleys beyond it, repeating perhaps the same lush system in perpetuity, out to a blurry, inconclusive horizon.

I think of this gaze as godlike not because of its grandeur, but rather because of the catastrophic and unstoppable force with which it is entangled. I am uneasy because here in the twenty-first century, I am a Black woman, of Caribbean descent, and an ecologist concerned with the origins of our current planetary crisis. The people whose lives and labor were currency within this very landscape are my people, my ancestors. The gaze of messengers like Church, whose art relayed the richness of nature for populations waiting for its dismantled parts, holds stories of both Black and ecological peril because of the truths they do not tell. The story told by the light shining across the leaves and the river as it winds through the valley is one of worlds cracked open, economies constructed, and human lives renegotiated. Observing this gaze, I see the all-too-common European desire to rewrite reality, the creation of dense tropical forests where centuries of sugarcane production through slavery had reshaped Jamaica's hydrology, leaving behind a drought at the time the painting was created.

Islands such as Jamaica were and are the simultaneous hosts and feasts for the powerful delusion of European superiority. This concept, insidious in society to the point of invisibility now, is made clear to the patient observer of ecosystems. Nature betrays its captors and speaks for those who have been silenced in the

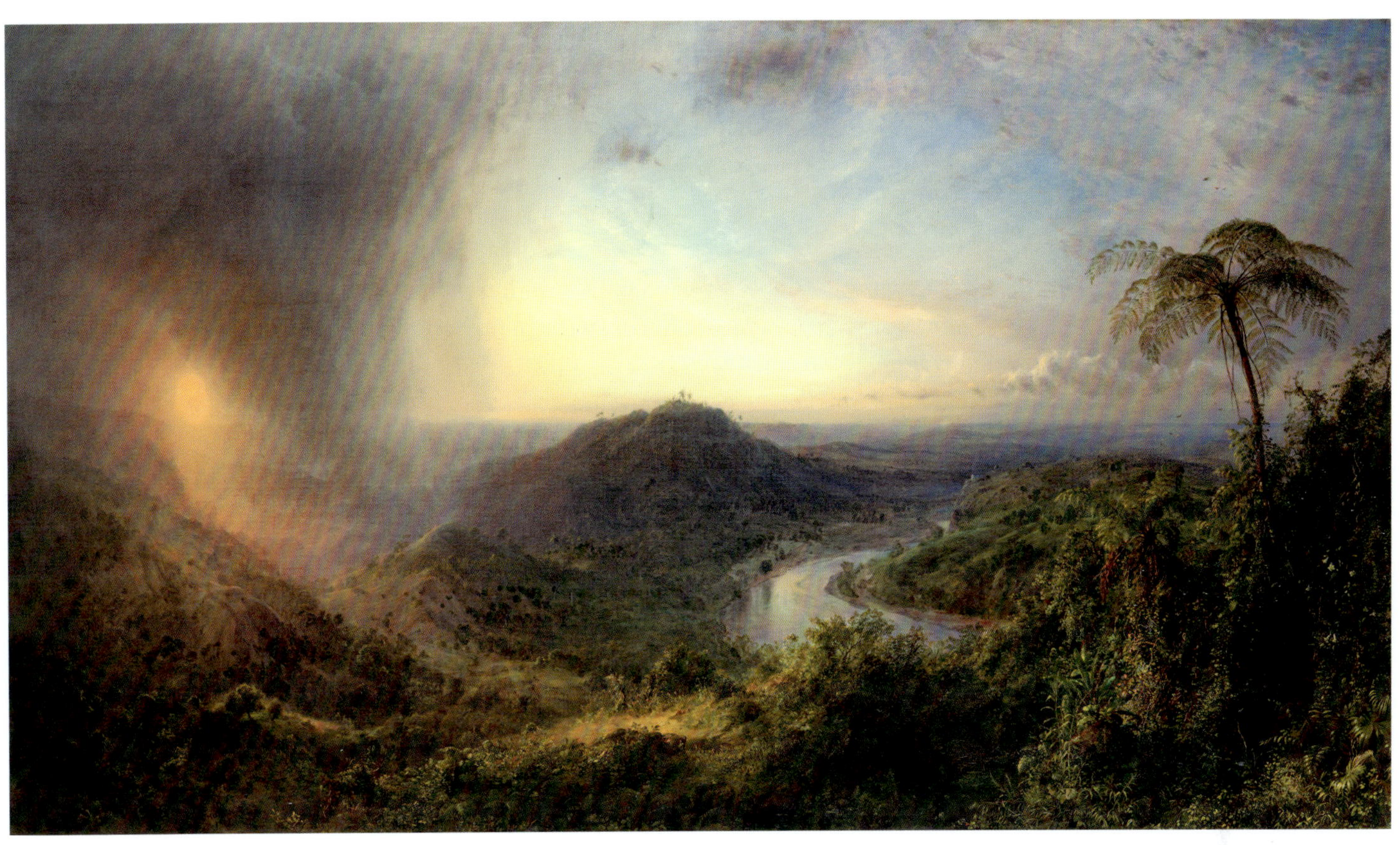

Frederic Edwin Church, *Vale of St. Thomas, Jamaica*, 1867. Oil on canvas, 48⅜ × 84⅝ in. (122.7 × 215 cm). Wadsworth Atheneum Museum of Art, Hartford, Connecticut, The Elizabeth Hart Jarvis Colt Collection, 1905.21.

Inspired by Alexander von Humboldt's descriptions of tropical nature, Church visited Jamaica in 1865. His paintings proclaim its landscape an untouched paradise, seen in this sweeping vista across a lush valley. However, the dry, sparse hills at left (nearly obscured by a dramatic rainstorm) reveal that Saint Thomas Parish was experiencing a severe drought, exacerbated by deforestation resulting from a history of plantation agriculture. Church also leaves out the homes and cultivated plots of the valley's residents, formerly enslaved people whose drought-induced suffering and government mistreatment ignited a brutally repressed rebellion within months of the artist's visit.

course of its disturbance. In the calculus of White, European supremacy above all other beings, human or otherwise, the *Vale of St. Thomas* is made synonymous with those Black bodies who are rendered "nonbeings" or "geologic life."[1] An imagination that thrives upon this flawed conceit organizes the ecosystem of a place like Saint Thomas Parish, Jamaica, into resources to extract, sell, and trade, at the price of the life of another human being. Looking out across the "vale," could Church see the exchange rate in timber, fertile soil, or gold for one enslaved African person? To the colonizing mind, is every new paradise also an accountant's ledger? The vocation of a landscape painter such as Church propagates the very desirable, even palliative, notion that the colonial enterprise that White supremacy operates can coexist with the version of Jamaica rendered in the *Vale of St. Thomas*—resplendent, glowing, and absent of Black life. Here in the present, the *Vale of St. Thomas* offers a visceral reminder that the beauty of nature can stop us from looking closely, methodically, and recognizing it as a site of Black struggles for personhood.

Suzanne Pierre is an ecosystems ecologist and biogeochemist focused on plant–microbe interactions. She is the founder and lead investigator of the nonprofit Critical Ecology Lab.

Dennis Carr

Thomas Cole's *Portage Falls on the Genesee*

In August of 1839, the British-born American artist Thomas Cole traveled to western New York to sketch a bend in the dramatic gorge along the Genesee River near Portage Falls. His surviving pencil sketch from that trip (shown on page 124) captures not only the natural rock formations, grandeur of the valley, and remnants of untouched wilderness, but also the encroachment of civilization—a house at the top of the cliff and the beginnings of a workers' camp above the falls. Cole had been hired to travel that summer by the commissioners of the canal system that was being built through the Genesee River Valley, an offshoot of the already successful Erie Canal. Ironically, this commission was made to document the wilderness before it became irrevocably altered by the construction project. Cole, whom the commissioner of the painting, Samuel B. Ruggles, described as "our 'American Claude'" (a reference to the French landscape painter Claude Lorrain), was the ideal artist for the job—a careful observer of nature, an environmentalist poet, and one of the most renowned painters of the American landscape.[1]

The resulting canvas (on page 125), which Cole likely completed later that fall back at his Catskill studio, is filled with his characteristic autumnal hues. It is also one of the largest canvases he ever painted—a massive vertical composition measuring seven feet tall by five feet wide. Cole, who traveled back to his native Britain to study the works of fellow artists, had recently returned from seeing John Constable's famous "six-footer" landscape paintings in 1829–32, an example of which is The Huntington's *View on the Stour near Dedham* from 1822 (see fig. 1.7).[2] Cole flipped his composition vertically, in order to capture the natural drama of the gorge, and he one-upped Constable by adding an extra foot to the canvas in each direction. Cole also added a dramatic storm cloud at the upper right, signaling the encroachment of civilization, an ominous portent of change. At lower right are his signature blasted trees, in this instance two gnarled trees perched on a rocky outcropping, resting across from each other as though having met some mutual fate.

Cole has also placed himself in the composition, in the lower left, having descended down the cliff—his small size dwarfed by the sublime majesty of the canyon (see page 126). Cole's inclusion in the painting, sketching with pencil in

Thomas Cole, *On the Genesee*, 1839. Graphite pencil on off-white wove paper, sheet: 14⅜ × 10¼ in. (36.5 × 26 cm). Detroit Institute of Arts, Founders Society Purchase, William H. Murphy Fund, 39.191.

Thomas Cole, *Portage Falls on the Genesee*, c. 1839. Oil on canvas, 84¼ × 61 ¼ in. (214 × 155.6 cm). The Huntington, Gift of The Ahmanson Foundation, 2021.8.

Cole's monumental painting depicts the Genesee River Valley in upstate New York, evoking the vast depth of the gorge through the canvas's upright format, and pinpointing the time of year with a palette exploding with the oranges and yellows that characterize autumn in the region. Though his image emphasizes the grandeur of nature, the landscape Cole depicts was under threat from industrialization. Construction for a new canal along the river was about to commence, as the workers' camp depicted just above the falls indicates. The roiling gray clouds at upper right and blasted tree at lower right perhaps suggest an awareness of humanity's destructive power.

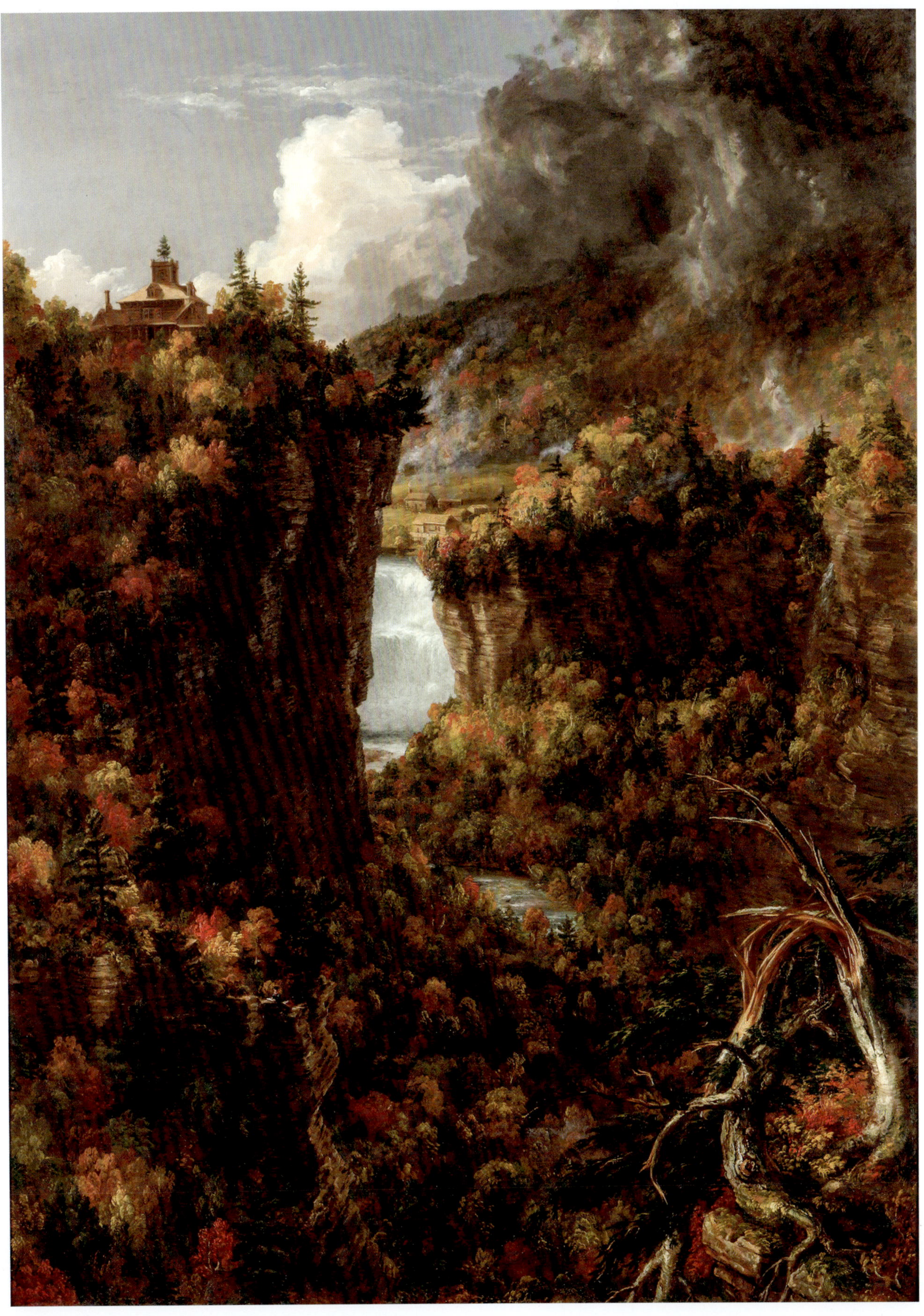

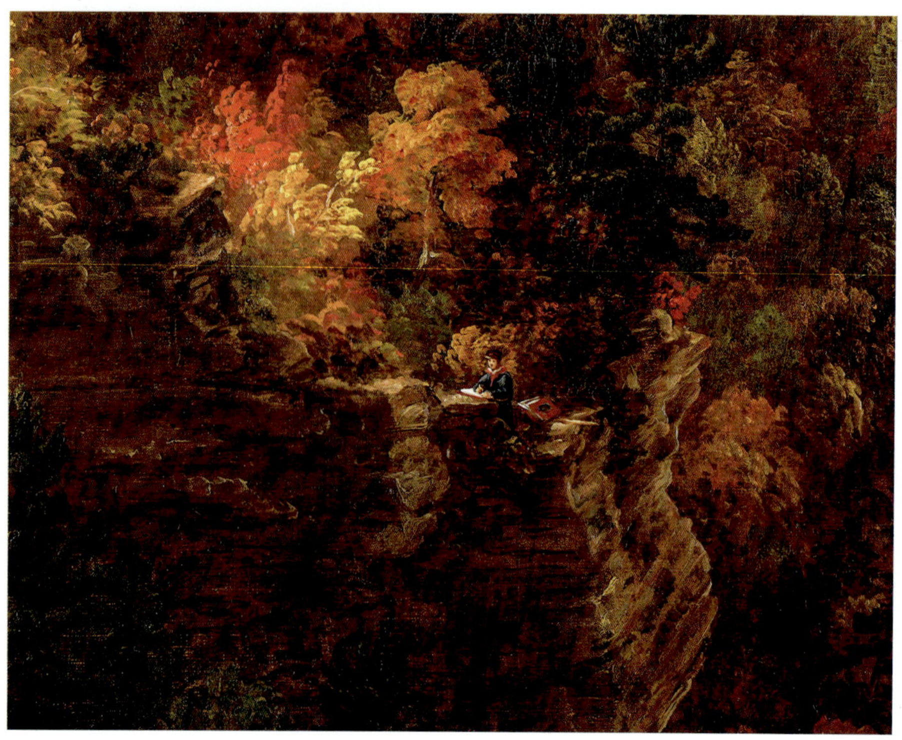

Detail of Thomas Cole, *Portage Falls on the Genesee*, c. 1839.

hand and surrounded by his portfolios, is reminiscent of his other large-scale canvas, *View from Mount Holyoke, Northampton, Massachusetts, after a Thunderstorm—The Oxbow* (1836, The Metropolitan Museum of Art), which shows the painter in the landscape straddling wilderness and civilization. Like *The Oxbow, Portage Falls on the Genesee* has a dramatic, divided composition, with wild nature on one side and encroaching civilization on the other, with Cole himself positioned as firsthand observer and arbiter of environmental change. Engineering problems ultimately caused the builders to abandon this section of the canal, and thus the magnificent landscape documented in Cole's painting was partially preserved, now Letchworth State Park. The painting's iconic connection with the landscape of New York State was reinforced when Ruggles presented it to then-governor William H. Seward to hang in the Executive Chamber in Albany. It remained in Seward's possession until his death in 1872.

Dennis Carr is the Virginia Steele Scott Chief Curator of American Art at The Huntington Library, Art Museum, and Botanical Gardens.

Richard Primack

Thoreau Speaks to Us about Climate Change

> Let us spend one day as deliberately as Nature, and not be thrown off the track. . . . Let us settle ourselves, and work and wedge our feet downward through the mud and slush of opinion, and prejudice, and tradition, and delusion, and appearance . . . to a hard bottom, which we can call reality.
>
> —Henry David Thoreau, *Walden* (1854)

In 1845, the twenty-seven-year-old environmental philosopher Henry David Thoreau built a small cabin in the woods by Walden Pond in Concord, Massachusetts. For two years and two months he lived there, recording nature observations and insights in his journal, and distilling them into the immensely influential book *Walden* (see page 128). In 1851, Thoreau began another nature observation project, this one an eight-year effort walking throughout Concord for several hours each day, recording the blossoming times of wildflowers, the leafing out of trees, and the first arrival dates of migratory birds.

In *Walden*, Thoreau coined the word "realometer," meaning an imaginary instrument that can measure reality. This instrument would allow people to push past the distractions that result from repetitive, mindless work and instead measure the essence of things. Thoreau's close observations of nature during his daily walks were crucial to his experience of reality.

Thoreau's 1850s observations have served as a modern realometer in ways he could not have anticipated. In 2003, my students and I began repeating Thoreau's walks, recording when plants flowered in many of the same locations Thoreau saw them. We found other people—past and present—who also recorded when they saw wildflowers, leaves, birds, and ice around Concord, and pulled together their data from field journals and personal records into spreadsheets. When we combined our own observations with Thoreau's and those made by other Concord residents, we found incredibly strong evidence that climate change is altering the natural world. Plants are flowering earlier in the spring and vulnerable wildflowers are declining.

This type of observation of nature gives each of us our own realometer—we can see and record the real-time effects of a changing climate on the world

"The [illegible] circle, wherein man and they act in cooperation,
That thou mayest get thy daily bread and not eat it in indifference;
revolve for thy sake, and are obedient to command; it must
be an effect of equitable condition, that thou shouldst be obedient too. [illegible]"
Saadi's Gulistan.

p. 8

Walden,

or

Life in the Woods.

Addressed to my Townsmen.

By

Henry D Thoreau.

~~At the time~~ When I wrote the following pages ~~were written~~ I lived alone in the woods, a mile from any neighbor, in a house of my own building, on the shore of Walden pond in Concord Massachusetts, and earned my living by the labor of my hands ~~exclusively~~ only. I lived there two years and two months. At present I am a sojourner in civilized life again.

I should not obtrude myself and my affairs so much on

around us. In my own case, after twenty-one years of observing flowering times of plants around Walden Pond, I see for myself that highbush blueberries are flowering two to three weeks earlier than in Thoreau's time, and that the ice in Walden Pond melts earlier in the spring than Thoreau ever recorded it.

A growing number of Concord residents are now making these observations, too, including recording when birds arrive in the spring, allowing them to determine which species are tracking climate changes and which are not. Each observer can see the reality of changes taking place in the forests and meadows where they live and recreate.

As we see Concord shift, it is natural to ask what we can do to stop or slow the effects of the climate crisis. I try to follow in Thoreau's footsteps and observe nature, live simply, and argue for needed transformations in society. Most importantly, reducing our use of natural resources, especially fossil fuels, both as individuals and as nations, can minimize the greenhouse gases that cause the climate crisis. Remaining aware of our environmental realometers can keep us on the right track.

Richard Primack is a professor of biology at Boston University. He is the author of Walden Warming: Climate Change Comes to Thoreau's Woods *(2014).*

Title page of Draft B-C of Henry David Thoreau's *Walden; or, Life in the Woods*, autograph manuscript, 1848–54. The Huntington, mssHM 924.

Henry David Thoreau began writing in his journal while living on Walden Pond in Massachusetts between 1845 and 1847. Over the next few years, he expanded these journal entries into an increasingly long essay about his time there. This was eventually published as *Walden; or, Life in the Woods*, a meditation on the importance of remaining attuned to nature in an increasingly industrializing world. The book was not an immediate financial or critical success when published in 1854 but has become one of the classics of American literature and a model for many generations of naturalist writers. The Huntington's Library holds Thoreau's manuscript drafts of *Walden*.

Kim Stanley Robinson

John Muir in 2024

By the time John Muir first walked up into the Sierra Nevada of California in 1868, he was thirty years old, and by no means an innocent. He had led a very full life, including a carefree childhood in Scotland, ten years as a farmhand toiling for his stupidly abusive father, a good education at the University of Wisconsin, a starvation-induced mystical ecstasy on the north shore of Lake Ontario, three years as a draft dodger in Canada, a stint in an Indiana machine shop (including an accident that blinded him for a month, with no guarantee sight would return—this marking him forever), a long walk across the post–Civil War South, and a near death from malaria. All these intensities combined to make him supremely ready for his encounter with the High Sierra, and it struck him with immense force, as an unexpected and exhilarating transport to a higher world, a heaven in this world and this life. He fell in love, and that love never left him.

His book *My First Summer in the Sierra* (shown on the facing page) comprises his journals from that year, and as such gives a vivid portrait of his contemporary feelings. His later reflections on his Sierra start are interesting, and often rapturous, but they necessarily lack the electrifying sense of surprise that often struck him in those first months of discovery.

His enthusiastic writing about the Sierra helped spark a worldwide shift in humanity's perception of Earth's biosphere, from one of instrumentalist extraction to ecological balance. If problems were created by some of the changes

---

John Muir, *My First Summer in the Sierra* (Boston: Houghton Mifflin, 1911). The Huntington, 20313.

Naturalist and author John Muir was an early advocate for the preservation of wild landscapes in the American West. *My First Summer in the Sierra* details the two years he spent living in a small cabin he built in the Yosemite Valley soon after arriving in California. Concerned over damage done to its meadows by domestic livestock, such as sheep, and the logging of its giant sequoias, Muir sought the same protections for Yosemite as those given to Yellowstone when it was established as the United States' first national park in 1872. As a result of Muir's efforts, Congress created Yosemite National Park in 1890.

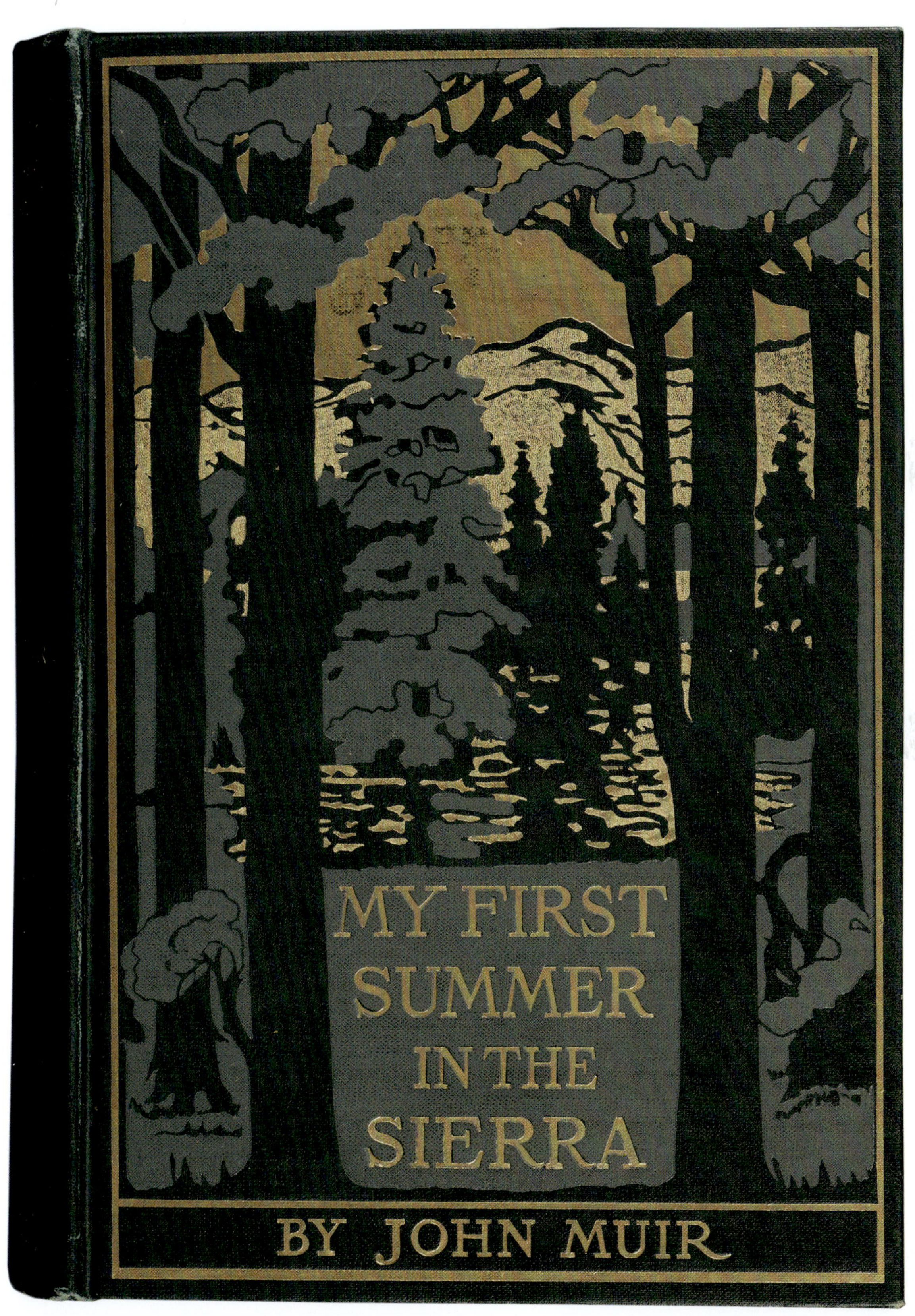
MY FIRST
SUMMER
IN THE
SIERRA
BY JOHN MUIR

resulting from this shift of consciousness, they were not of his making, and his fundamental intuition about our interconnectedness with all living creatures on Earth was crucial in its time, and is needed now more than ever. Though Muir will always read like "a white man [writing] in the nineteenth century" and has recently been criticized for some aspects of that worldview, that writing includes some of the most beautiful, prescient, and important thoughts ever expressed about our relationship to our planet and to our fellow creatures.[1] For me he is still and always will be a vivid writer, a charismatic movement leader, a true mountain lover, and the most famous environmentalist of all time. His advocacy for animals both domestic and wild, "our horizontal brothers and sisters," as he called them, and his lifelong work to preserve wild spaces, and to encourage others to do the same, will always be a glorious intervention in world history, and an unfinished project for us to pick up and pursue.

His childhood home in Dunbar, Scotland, is a museum to his memory. His adult home in Martinez, California, is a National Historic Site. There are signs in Wisconsin near the farm where he worked through his teens, in Canada where he sat out the war, and in Indianapolis where that shard of metal shot into his eye. His real home, the High Sierra, is there for all to visit and enjoy, much the same as when he first walked up into its glory. He inspired the preservation of that wilderness for the generations to come, and for that I will always be grateful to him and honor his memory.

Kim Stanley Robinson's novels include the Nebula Award–winning Mars *trilogy (1992–99) and* The Ministry of the Future *(2020), about climate crisis. His most recent book was* The High Sierra: A Love Story *(2022).*

Alan Bacock

A *Paya* (Water) Story

Los Angeles. The name conjures images of beaches, palm trees, sports teams, and celebrities living in lavish mansions. Today it is a city of 3.8 million people known globally as the entertainment capital of the world. However, the wealth of Los Angeles cannot be compared to the riches that the "land of little rain" has provided. In fact, Los Angeles owes its very existence to this remote land that had its resources diverted to Southern California.

The region is situated within one of the deepest valleys in the United States, about 250 miles northeast of Los Angeles. Located between the towering Sierra Nevada mountain range to the west and the Inyo-White mountain ranges to the east, the deepest valley has little chance of precipitation due to the rain shadow caused by the Sierra, which shelters the land from rain-bearing clouds. Mary Hunter Austin called this landscape the Land of Little Rain in her book of that title (shown on page 134).

What does this parched place hundreds of miles from Los Angeles provide the city? Water. It is true that little rain falls to the ground here, but water has been an abundant part of the story of this place.

My people, who have called this place home since time immemorial, named this valley Payahuunadü, which can be translated as "the place where water flows." The snow that falls and rests on the Sierra Nevada throughout the winter melts as the days become hotter, filling up waterways leading down into the valley below. As the meltwater moved to the lower elevations, my ancestors spread these waters by developing a complex network of ditches. Slowing the water and spreading it allowed life to flourish.

All of this changed when ranchers and miners entered the land, implementing their own set of rules based on human ownership. My ancestors lived in community with all organisms and struggled to understand how these laws for land, water, and animals were beneficial for the whole of creation. In the first decade of the twentieth century, fifty years after the land and water were removed from our influence, new foreigners from Los Angeles used those same laws to take away the land and water from those who first took from us. However, this time it meant that our riches would not remain in Payahuunadü but would be channeled and diverted to the big city.

The Los Angeles Aqueduct opened in 1913. It is celebrated as an engineering marvel, moving water 250 miles through deserts and over mountains while

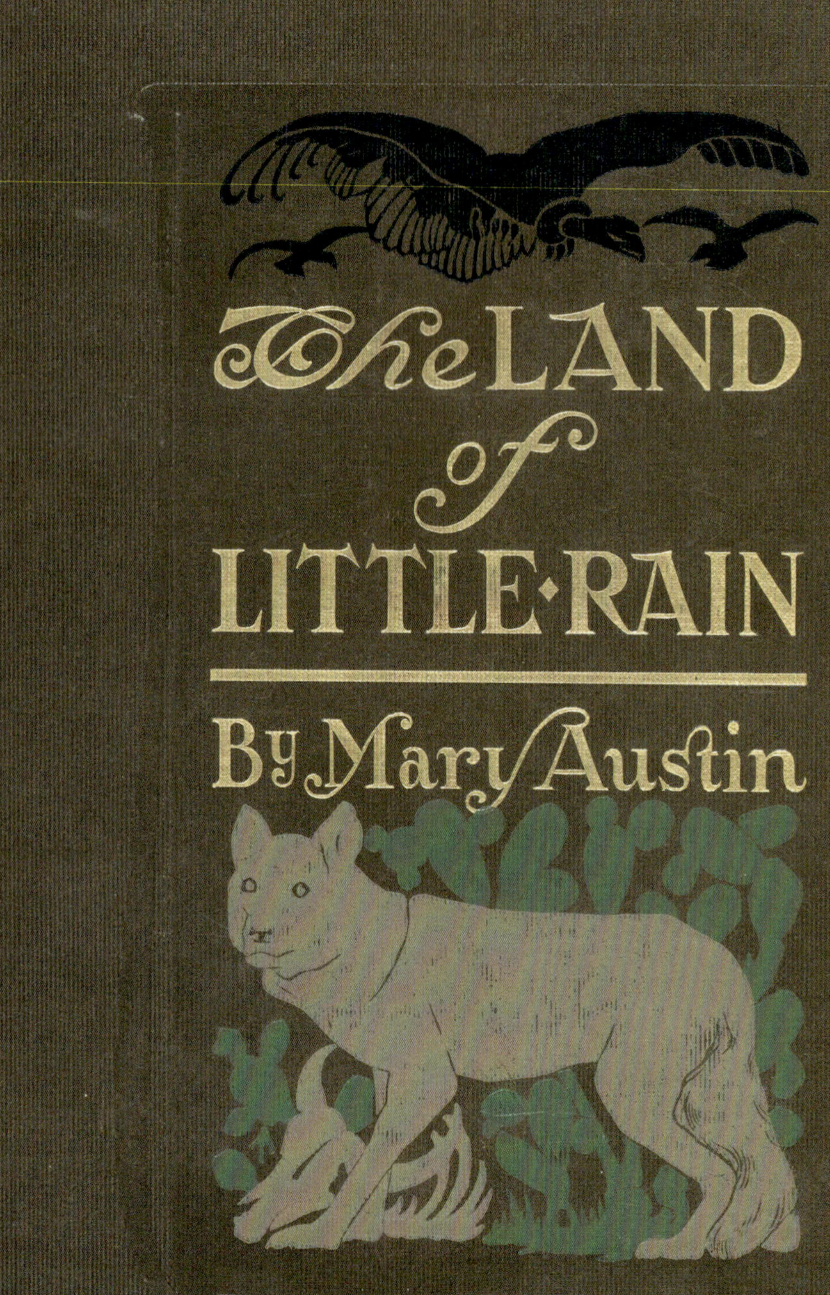
The LAND
of
LITTLE·RAIN
By Mary Austin

adding electricity to the grid. Great as it might seem from an engineering standpoint, the aqueduct was, in reality, built solely for human desires and uses. Those uses are not necessarily incompatible with the needs of nature, but the human-centered focus decided which need would be prioritized. For instance, Los Angeles desired Owens River water, which resulted in the draining of Owens Lake, which had devastating environmental impacts. In 2014, California passed a law to support more sustainable use of groundwater throughout the state. However, a last-minute addition to the law excluded the Owens Valley from its protections because lawyers determine what is "right" based on their clients' needs.[1]

My ancestors recognized the wonders of the local "eco-nomy" and it was not based on money. They lived within the beauty of Payahuunadü and learned how to survive in both abundance and scarcity. We would be wise to learn from Indigenous cultures built on knowledge handed down for thousands of years rather than rely on engineering marvels that are comparatively limited in their viability and may come at a cost that we have yet to fully comprehend.

Alan Bacock is a tribal member and former tribal council secretary of the Big Pine Paiute Tribe of the Owens Valley. He currently works for the federal government supporting the environmental protection goals of tribal communities in the Southwest.

Mary Austin, *The Land of Little Rain* (Boston: Houghton Mifflin, 1904). The Huntington, 642.

As a writer, Austin is particularly associated with the region of Southern California located between the High Sierra and the Mojave Desert. Her book *The Land of Little Rain* chronicles the plants, animals, and people found there—from coyotes and rabbits seeking water, to Indigenous craftspeople using natural resources, to settler mining communities enduring hardships in a place they love. Each chapter describes how these denizens of the harsh environment survive there and is critical of those who refuse to live in harmony with nature. Though her text romanticizes the area's Native people, Austin was an advocate for Indigenous rights and the desert landscape.

Nicole Cavender

Awakening to the Anthropocene

Understanding Soil, a Critical Resource

When we see the fruits of our labor, such as these peaches (on the facing page), it may not be obvious at first that their production depends on thousands of years of soil formation. Soil is a direct link to food security and sustainability. Soil is alive; soil is energy; all terrestrial life depends on soil in some capacity. Scientists believe that one tablespoon of soil contains a higher number of living organisms than there are people on this planet. Through a long, complex history of geological, climatic, and biological inputs, the creation of soil is an intricate process. It can take hundreds of years to form the first inch of topsoil.

Soil is essential for protecting natural resources, managing environmental systems effectively, and sustaining agricultural production to ensure global food security. Further, soil is an important part of nutrient cycles—such as the carbon and nitrogen cycles. It is critical for filtering rainwater, preventing flooding, anchoring the roots of plants, and providing crucial habitat for insects, microbes, nematodes, fungi, bacteria, and many other organisms. Agriculture practices that use regular tilling and chemical applications such as fertilizer kill soil life and cause a drastic decline in long-term productivity.

While humans have farmed for millennia, advancements such as the plow, and the invention of the gasoline-powered tractor in the late 1800s, dramatically changed how we produce food and relate to the land. Industrial practices caused erosion and nutrient depletion. Large monocultures, or fields of the same plant, made food production more efficient and scalable, but this practice resulted in a loss of biodiversity, above and below the ground. Poor land leads to economically insecure people and undermines the ecology that we depend on for survival.

Around 95 percent of the food we eat comes from the soil, directly or indirectly, and this especially impacts residents of California, the country's top agriculture producer and exporter. Of California's roughly one hundred million acres, more than a quarter is currently being used for cropland. Throughout California's history, its soil has been an important resource, and Carleton Watkins's album was commissioned as an advertisement. This changing relationship is on full view in Watkins's images of agriculture in California's Kern County from 1881 to 1889.

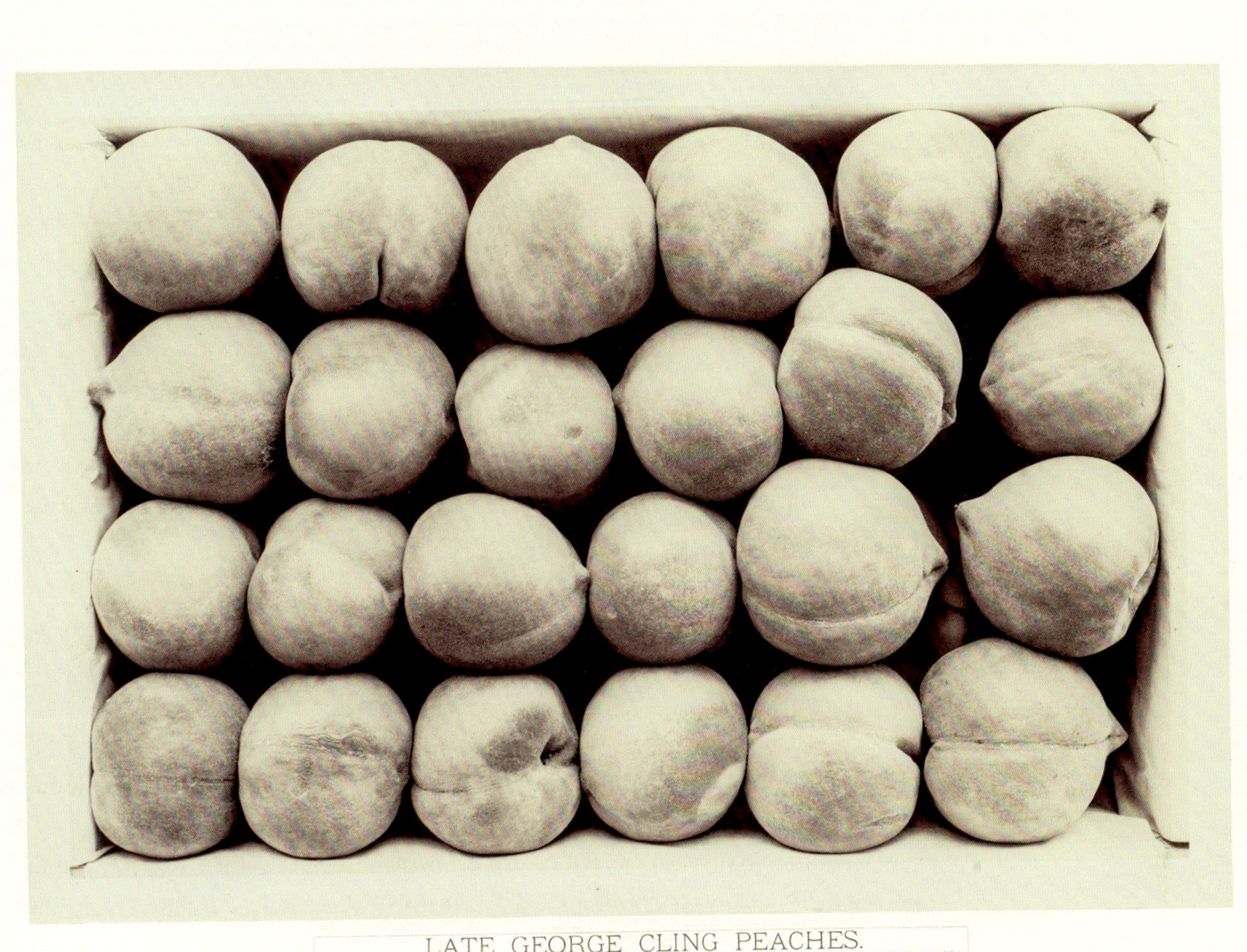

"Late George Cling Peaches," albumen print [19 × 25 in. (48.5 × 64 cm)] in Carleton Watkins's *Photographic Views of Kern County, California* (San Francisco: Carleton Watkins, c. 1881–89), plate 12. The Huntington, 137500.

Carleton Watkins created this album of monumental photographs for his clients, landowners in the San Joaquin Valley, to serve as evidence in a lawsuit concerning water rights. The images trace the "life cycle" of the region's crops, from the canals and irrigation ditches necessary to planting to a display of ripe produce at an exhibition. One photograph features a variety of peaches developed with shipping in mind. The contrast between the shadowy depths of the box and the bright, fuzzy peach skins emphasizes the fruit's round shape and suggests its lusciousness.

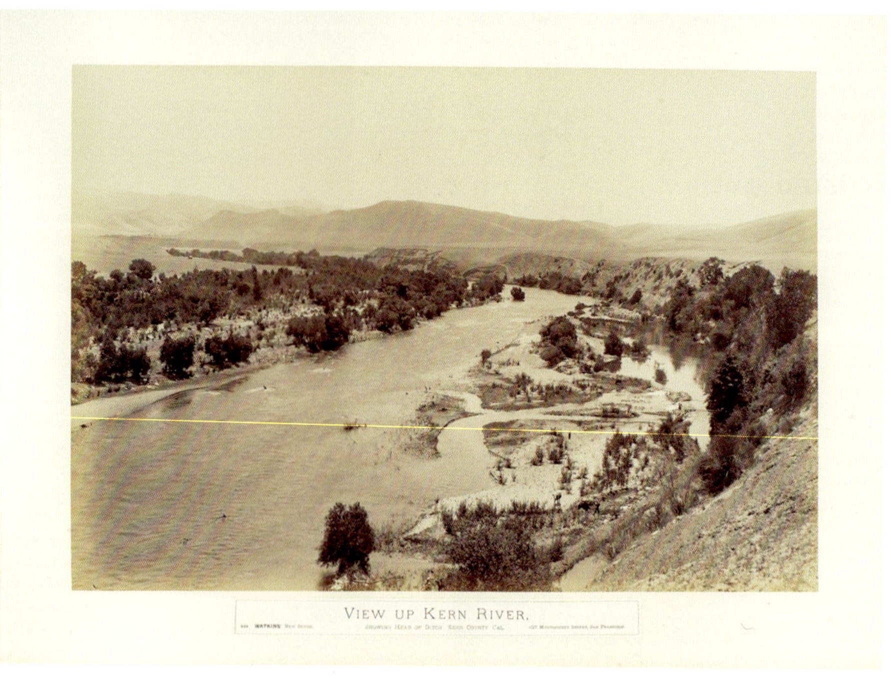

"View up Kern River, Showing Head of Ditch, Kern County, Cal.," albumen print [19 × 25 in. (48.5 × 64 cm)] in Carleton Watkins's *Photographic Views of Kern County, California* (San Francisco: Carleton Watkins, c. 1881–89), plate 25. The Huntington, 137500.

As we face the impacts of the environmental crisis, soil regeneration is a key solution because it is the greatest source of terrestrial carbon storage. Practices such as tilling, where surface plants are removed and topsoil is overturned, expose barren land and release carbon into the atmosphere. Furthermore, it causes topsoil to flow into the waterways due to erosion, creating situations like that seen in plate 25 of Watkins's album (shown above), where soil from barren banks washes into the Kern River. If the soil is protected by vegetation or cover crops, as opposed to continually being tilled, it stays in place, and it can become a carbon sink.

Our awakening to the Anthropocene can lead us to examine the past and reimagine the future. Regenerative agriculture requires a paradigm shift, but it will result in a regenerative economy. There are many actions we can take to support diverse and healthy soils. First is to stop the continual practice of tilling. Instead, farmers can integrate perennial plants and trees into their fields, as well as diversify the types of crops they plant. Holistic livestock management can support healthy crops and healthy soils. Nutrients and chemical applications can be applied and used more judiciously. Organic portions of the soil can be promoted and replenished through compost. Technology, ingenuity, and human responsibility can be harnessed to support these goals. Humans are going to continue to need soil to produce food, and the choices we make regarding this resource will determine the quality of the future for people and the planet.

Nicole Cavender is the Telleen/Jorgensen Director of the Botanical Gardens at The Huntington Library, Art Museum, and Botanical Gardens.

Kristen Anthony

The Promise of Smoke

Lazell, Perkins, and Company commissioned this advertisement (on page 141) around the middle of the nineteenth century. At the time, they were one of the largest ironworks in the United States, and prided themselves on their contribution to American enterprise.[1] The scene illustrates the American fantasy of industrial prosperity made possible by an allegedly endless supply of coal, lumber, and labor. There are no fewer than sixteen active chimneys, flues, and smokestacks. Pristine Georgian-style buildings sit uphill but downwind of these billows and remain improbably white. The overall effect of the image is not a grotesque scene of immoderate pollution. The clean, light blue sky composes almost half of the composition, signaling a promising future. The steam train in the foreground is rendered in bright colors as it chugs happily along. It is likely heading out west, where it will deposit its cargo, parts for "all kinds of machinery." The scene is undeniably optimistic, and in this way truly illustrative of the era.

It's difficult to overstate the sheer ubiquity of smokestacks in nineteenth-century visual culture. In conducting object research for this exhibition, I was astounded by the prevalence of this motif. I encountered smoke-vomiting factories on advertising posters, city-view prints, trade catalogues and cards, letterheads, stock certificates, and even product labels for companies of all types. These excrementitious plumes were not seen as sinister or undesirable, but rather taken as symbolic of industrial fortitude, power, and, above all, unprecedented productivity.[2] Rather than editing or omitting the polluting chimneys and smoke from these illustrations, companies were more likely to embellish—to add additional, fictional flues to the image to exaggerate their productive potential.

In industrial centers, these forests of flumes were so marvelous to spectators that they attracted tourism. In the 1910s, artist Joseph Pennell, while in Pittsburgh, exclaimed, "But how much more impressive is a row of blast furnaces, oil wells, and coal brakers than trees!"[3] Industry was indeed extraordinary; exponential manufacturing changed daily life for many by facilitating travel, communication, and increased access to goods. Yet, hand in hand with these growing industrial cities and towns came increasing complaints of respiratory illness, and factory labor was often exploitative and dangerous.[4]

Nonetheless, the image of the triumphant factory remained a mainstay in visual culture decades after these social and public health issues were well understood. Socialist artist Walter Crane displayed his understanding of these dangers when he appropriated the smokestack trope in his art, equating the image of industry with suffering. In one illustration (shown below) seated atop a smog-exhaling horse, the personification of capitalism shares a saddle with the figure of death, while factories loom in the background.

These once-celebratory tableaus are differently haunting to a present-day onlooker. The smokestack is now the symbol of climate change perpetration. Subsequently, scientists looking at two thousand years' worth of carbon emissions in ice core data would note that the upward trend occurs precisely when these images began to proliferate in print culture (see fig. 1.2). Although the emissions of the Great Acceleration of the mid-twentieth century far surpassed those of the previous century, the fossil-fueled train left the station during the industrial revolution. Despite the fact that many nations are working toward carbon reductions, the "fairytale of eternal economic growth" remains an insidious nineteenth-century legacy.[5]

Kristen Anthony served as the Assistant Curator for Storm Cloud: Picturing the Origins of Our Climate Crisis. *She holds a master's degree in history from California State University, Fullerton.*

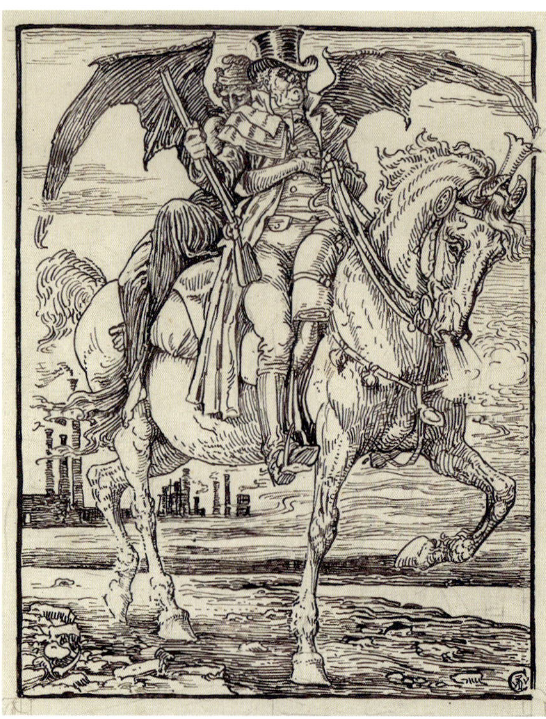

Walter Crane, *The Capitalist* (detail), c. 1889. Pen and black ink over graphite pencil on wove paper, sheet: 15 × 11 in. (38.1 × 27.9 cm). The Huntington, Gift of the Friends, 70.116D.

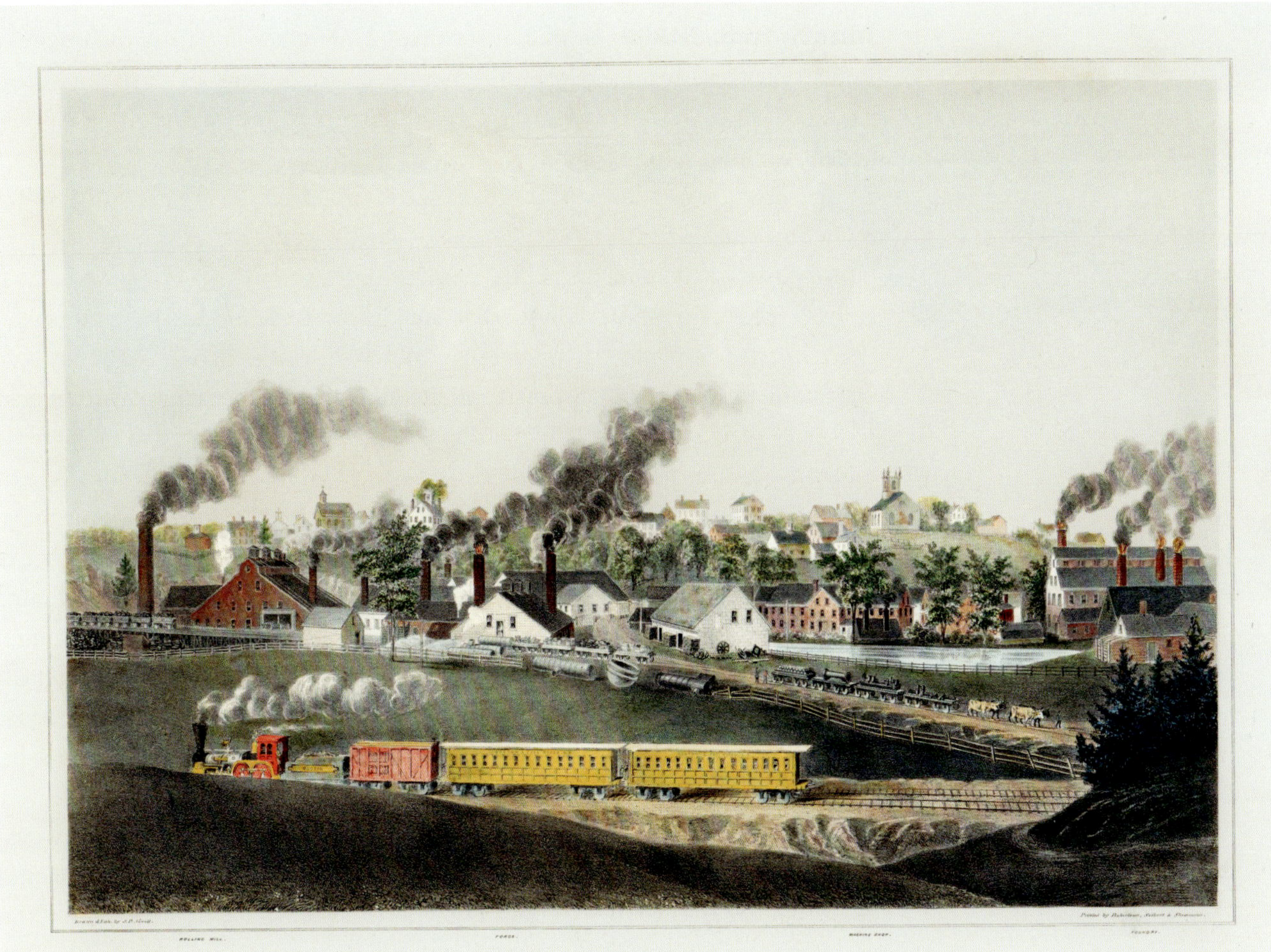

J. P. Newell (lithographer), *Lazell, Perkins & Co., Bridgewater, Mass.*, c. 1858–60. Color-printed lithograph with hand-coloring, 18⅝ × 23⅞ in. (47.3 × 60.6 cm). The Huntington, Jay T. Last Collection, priJLC 002553.

Once serving as advertisements for manufacturers, posters depicting factory scenes such as this were a common nineteenth-century marketing practice. Most factories at this time were powered by coal, the primary propellent of the industrial revolution and the dirtiest of all fossil fuels. Coal is the largest source of global temperature rise to date; it emits significantly more CO_2 than oil or natural gas. It also creates more particulate matter, hence the dark smoke. The consumption of coal increased exponentially throughout the nineteenth century, and was especially instrumental in iron production. Coal smoke is noxious, highly polluting, and linked to increased rates of respiratory illness.

Jonathan S. Blake

The Unlovely Forests of Los Angeles

> Suddenly the news had spread, in an explosion of excitement: an oil derrick!
>
> —Upton Sinclair, *Oil!* (1927)

Every time I step out my front door, I am confronted by the sight of pumpjacks atop a nearby ridge, lazily bobbing up and down to extract the liquified remains of organic matter from millions of years ago that rest deep beneath my home. The pumps sit one-half mile away, at the northern edge of the Inglewood Oil Field, one of the largest urban oil fields in the country. My family and I are just four of the one million people living within five miles of 672 active wells on 1,000 acres of Los Angeles.[1]

Yet oil drilling in LA today is nothing like it used to be. In the early twentieth century—when California was a global petro power, producing nearly one-quarter of the world's oil—"a forest of oil derricks" covered wide swathes of the city (see images on pages 144–47). Upton Sinclair detailed this period in *Oil!*, his 1927 novel inspired by the father of LA's oil industry, Edward Doheny (pictured beneath a derrick in the image on the facing page).[2] One grove in this vast urban forest tapped into the Venice–Del Rey Oil Field, where derricks dwarfed the seaside homes they abutted among the Venice canals. Oil was discovered there only six months before this photo was taken, in July 1930, but in that short time twenty-two companies received 180 drilling permits, with 25 more pending.[3] The oil bonanza went bust two years later, and eventually Venice returned to its first passion: real estate. A home on the canals is currently listed at $7 million.

Today, LA's extractive infrastructure is largely hidden from view (refining and storage are another story). Many of the region's sixty-eight named oil fields are still being drained, albeit less overtly, using underground operations and wells concealed inside unmarked, windowless buildings—including a 175-foot tower encasing forty wells tapping the Beverly Hills Oil Field on the corner of Pico Boulevard and, fittingly, Doheny Drive. However, the industry that helped transform LA from peripheral town to global megalopolis and fueled the icons of its self-image—the car, the freeway, the single-family home—is now being cast out. Los Angeles City and County banned all new wells and mandated that all existing oil and gas production cease by the early 2040s. California prohibited new oil wells near homes and schools, and, after 2035, the sale of new gas-powered cars. The fossil fuel industry is waging war on these measures, but I am hopeful, with Sinclair, that "some day all those unlovely derricks will be gone."[4]

After they are, and the land is cleansed of toxins, there's a plan to remake the Inglewood Oil Field into an enormous park. Perhaps one day the view from my doorstep will still be fields, not of oil, but of poppies and finches, and children "running over those hills."[5]

Jonathan S. Blake directs the Planetary Program at the Berggruen Institute. He is co-author of Children of a Modest Star: Planetary Thinking for an Age of Crises *(2024).*

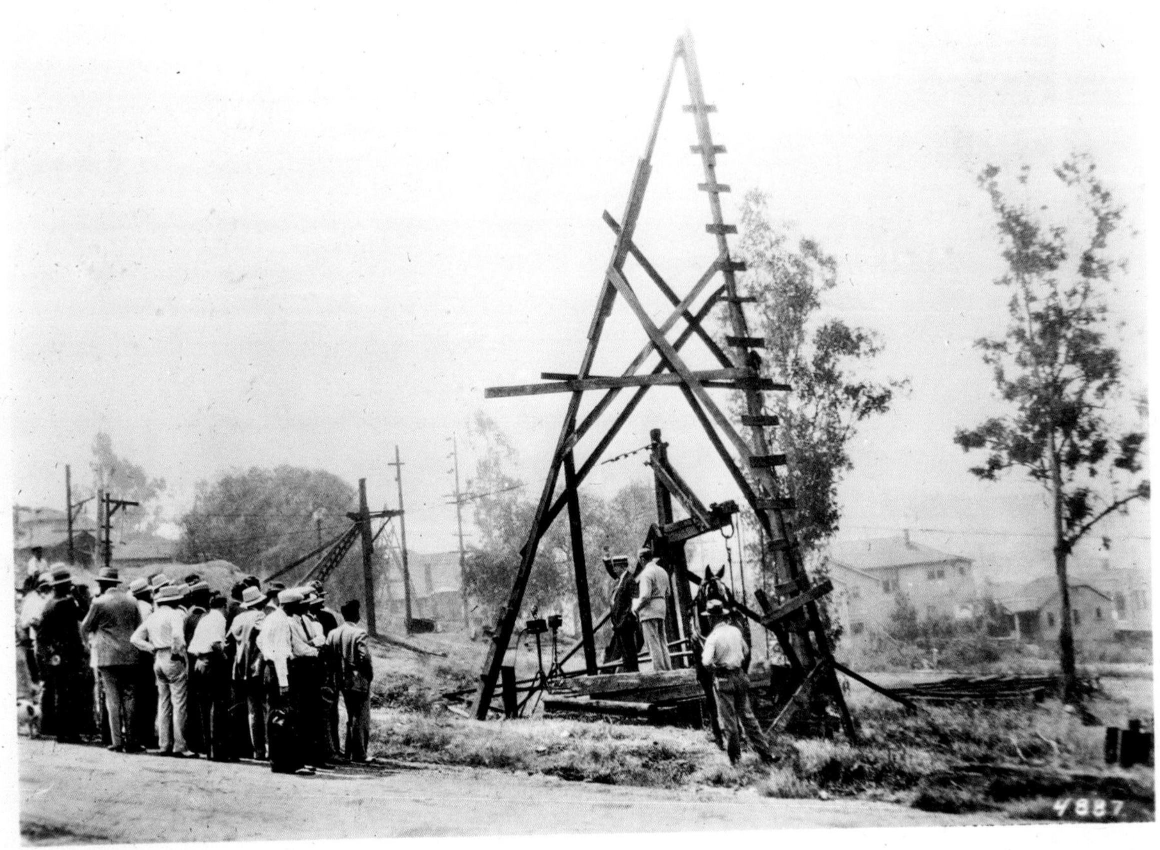

Charles C. Pierce, *Doheny Discovery Well Showing Mr. [Edward] Doheny with Raised Hand,* 1890. Gelatin silver print, 6¼ × 8¼ in. (16 × 21 cm). The Huntington, C. C. Pierce Collection of Photographs, photCL Pierce 06584.

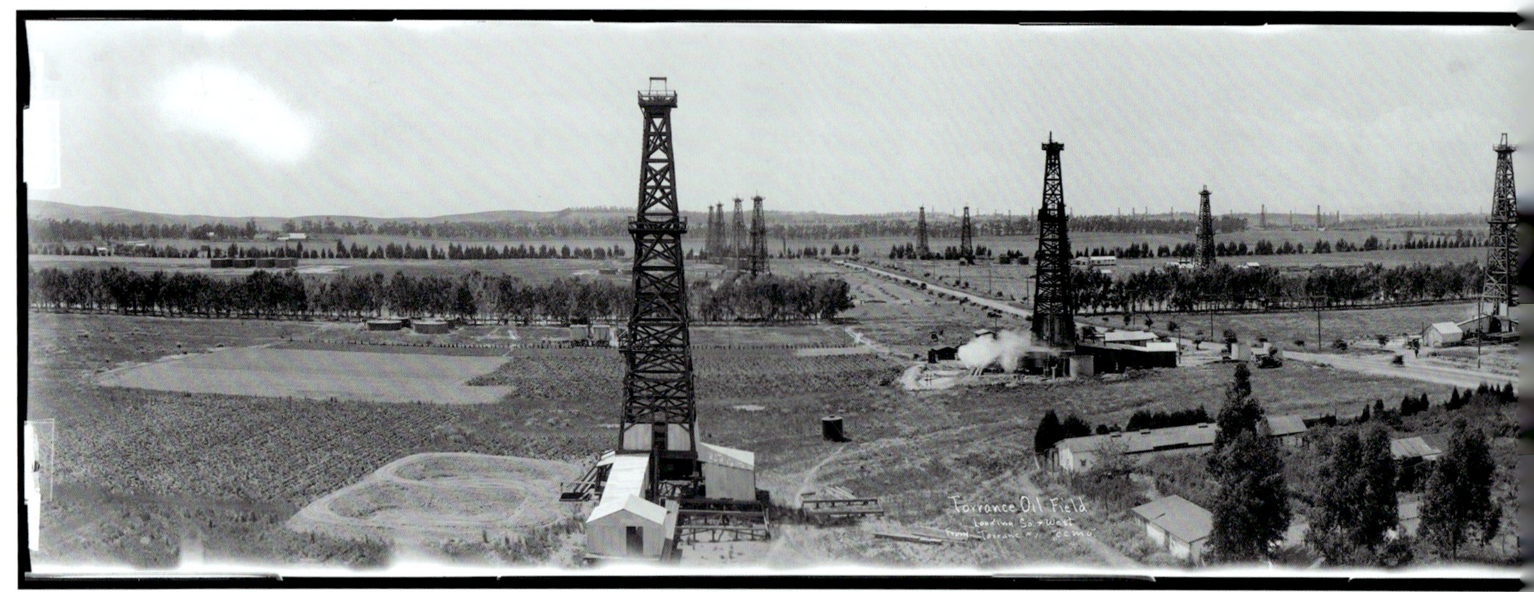
Torrance Oil Field
Looking So. West
From Torrance #1

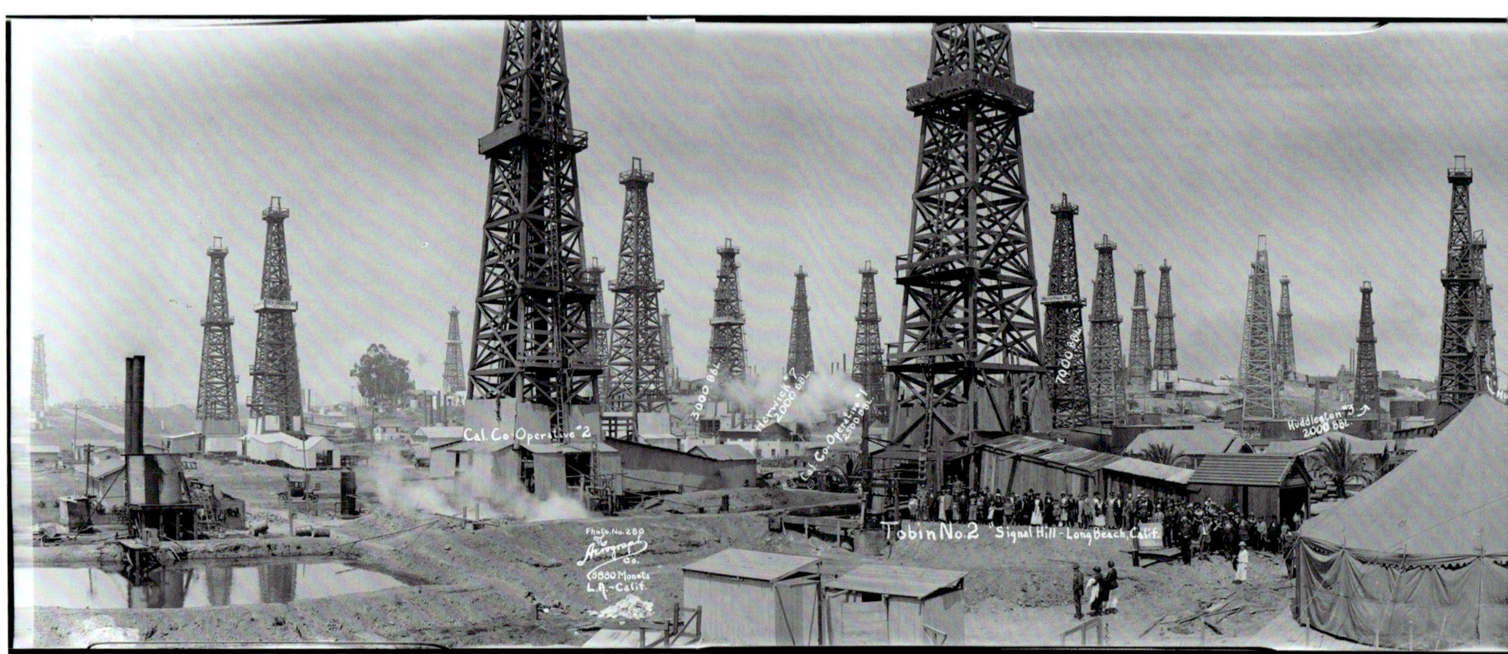
Cal. Co-Operative #2
3000 BBL.
Herwick #7
2000 BBL.
Cal. Co. Operative #1
2500 BBL.
7000 BBL.
Huddleston #3
2000 BBL.
Tobin No. 2 "Signal Hill - Long Beach, Calif.
Photo. No. 280
The Aerograph Co.
5880 Moneta
L.A.-Calif.

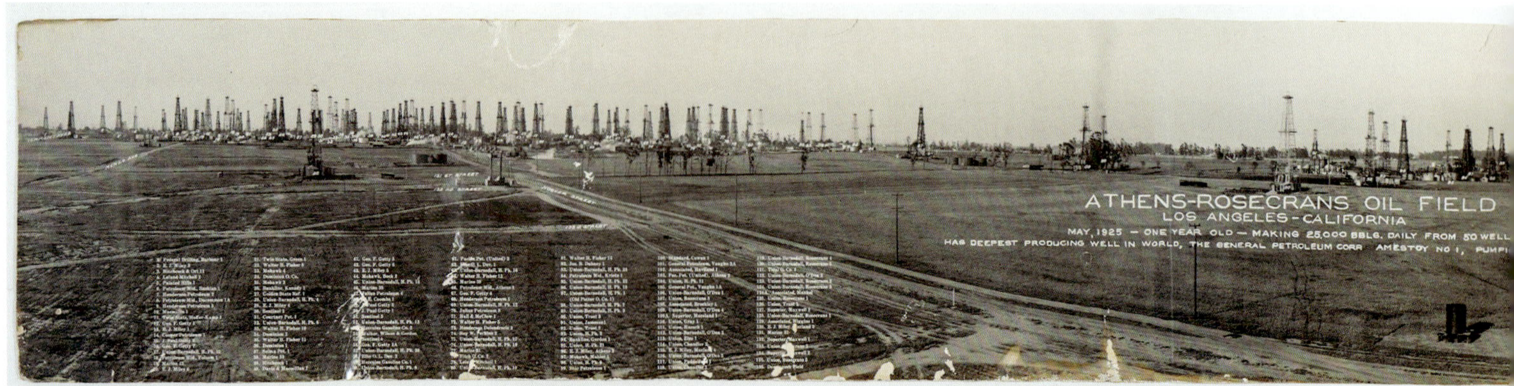
ATHENS-ROSECRANS OIL FIELD
LOS ANGELES - CALIFORNIA
MAY, 1925 — ONE YEAR OLD — MAKING 25,000 BBLS. DAILY FROM 50 WELL
HAS DEEPEST PRODUCING WELL IN WORLD, THE GENERAL PETROLEUM CORP. AMESTOY NO 1, PUMPI

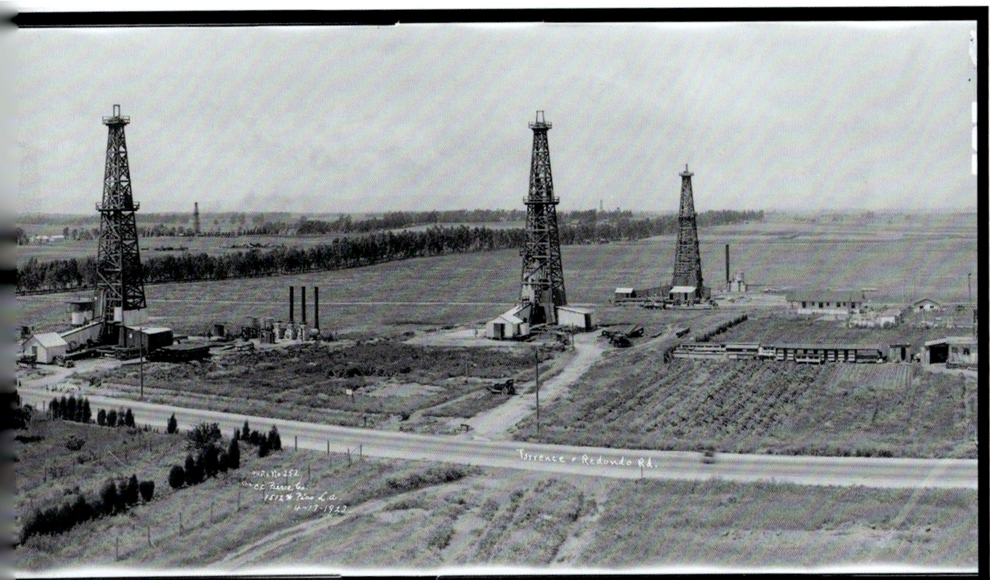

Charles C. Pierce, *Oil Field, Torrance, April 17, 1923*. Gelatin silver print, 10 × 46 in. (26 × 117 cm). The Huntington, Verner Collection of Panoramic Negatives, photCL 470 (088).

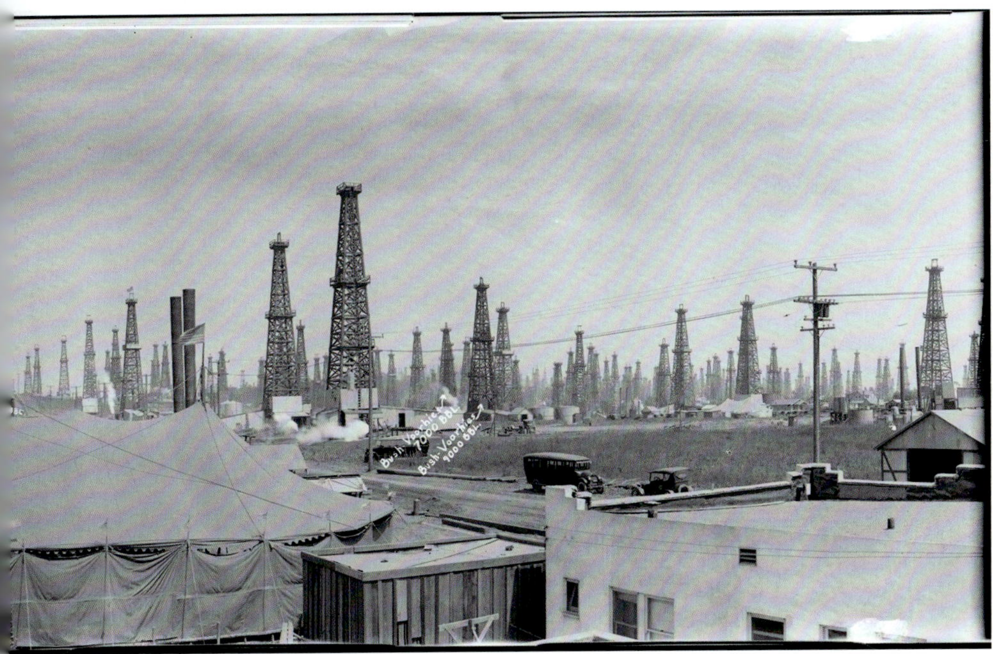

Aerograph Co., *Tobin #2, Signal Hill Oil Field*, 1923. Gelatin silver print, 10 × 42 in. (26 × 107 cm). The Huntington, Verner Collection of Panoramic Negatives, photCL 470 (172).

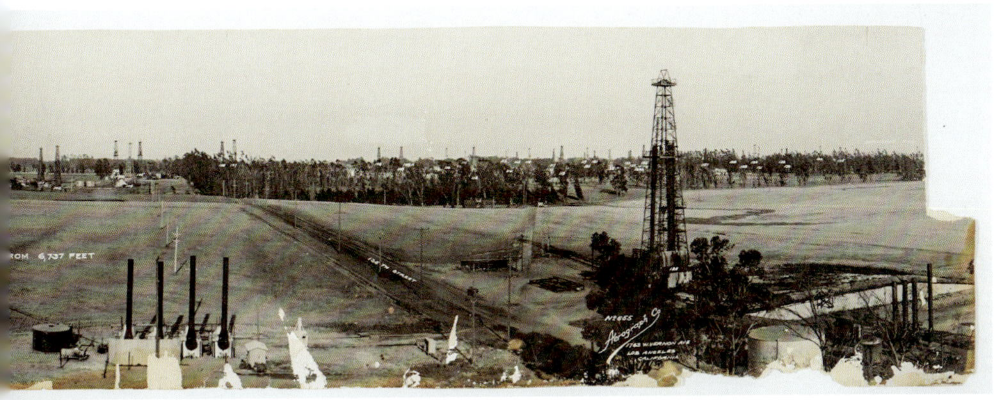

Aerograph Co., *Athens-Rosecrans Oil Field, Los Angeles, California*, 1925. Gelatin silver print, 8¾ × 62½ in. (22.2 × 158.8 cm). The Huntington, Ernest Marquez Collection, photCL_555_04_09.

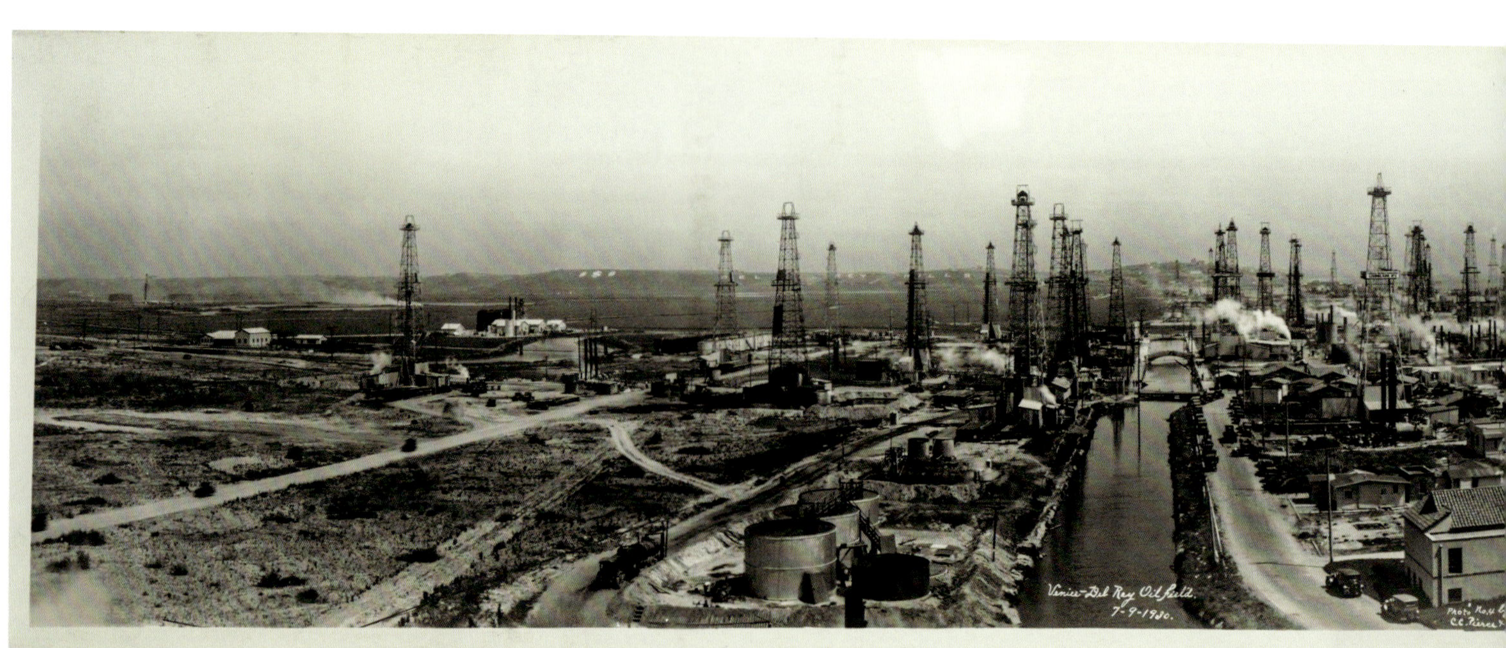
Venice-Del Rey Oil Field.
7-9-1930.
Photo No. 4
C.C. Pierce

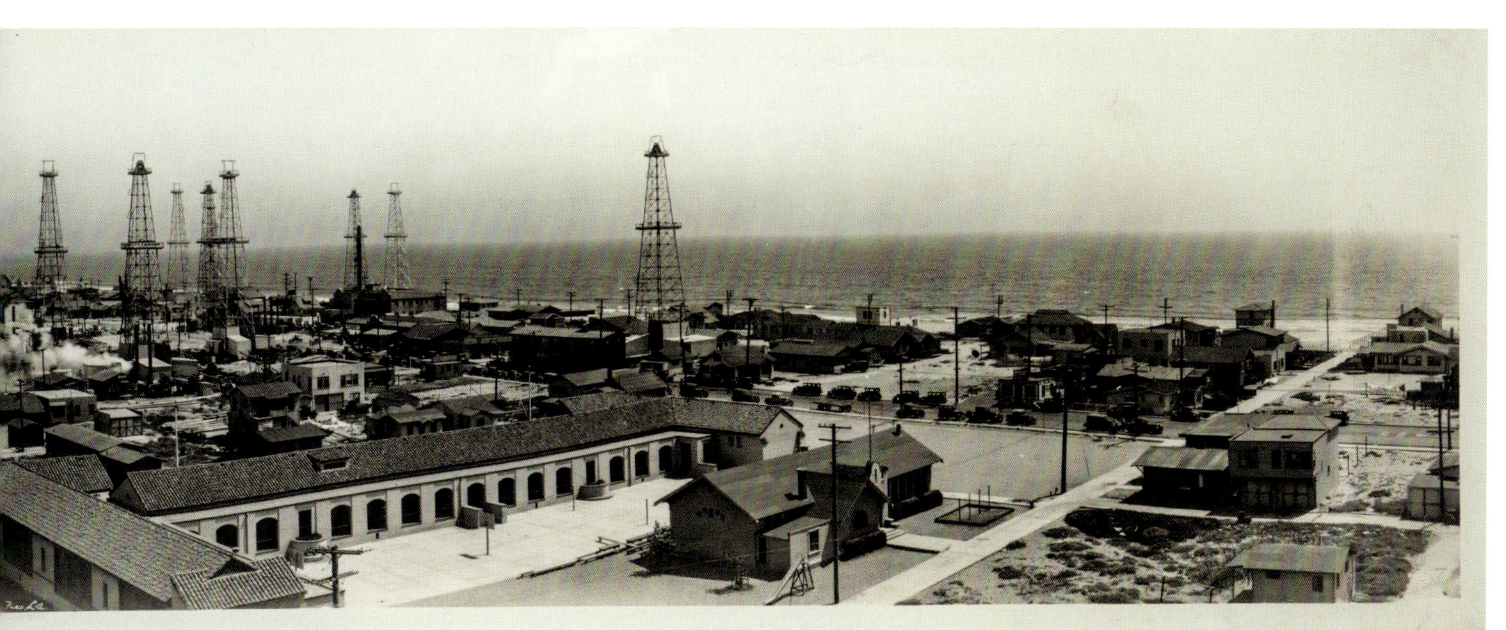

Charles C. Pierce, *Venice–Del Rey Oil Field, 7-9-1930*, 1930. Gelatin silver print, 9⅞ × 51 in. (25 × 129.5 cm). The Huntington, Ernest Marquez Collection, photCL_555_04_21.

Edward Doheny's 1892 discovery of an oil well in Los Angeles ushered in a drilling boom across the region, as these early twentieth-century photographs demonstrate. While most derricks have been dismantled, Los Angeles County remains the nation's largest urban oil field, with thousands of active wells pumping alongside a population of more than ten million people. Existing pumps are sometimes camouflaged behind fences, disguised as other structures, or, more often, placed in lower-income neighborhoods. In 2022, groups seeking environmental justice on behalf of these frontline communities successfully lobbied to ban new wells and phase out drilling within the city limits.

Rebeca Méndez

The Storm-Cloud of the Twenty-First Century

As an interdisciplinary artist working at the intersection of art and science, I examine reciprocal relationships and environmental justice in a multispecies world confronting climate change, mass extinctions, and a ravaging extractivist economy. I am constantly seeking a historical context for my research and art practice. There are parallels between John Ruskin's two lectures titled "The Storm-Cloud of the Nineteenth Century" (1884) and my work *Any-Instant-Whatever* (2020), as well as to my practice more broadly. A contemplation of the sky above Los Angeles, *Any-Instant-Whatever,* as the curator Stephen Nowlin writes, "portrays the daytime sky in its textures and physics of blueness, its range of water formations sculpted by wind and pressures born of our nearest star—forces which, when synaptically processed, transform into emergent sensations of sublime beauty."[1]

In "The Storm-Cloud," Ruskin describes precisely not only what one can visually perceive when viewing clouds but also "the way in which clouds, as a matter of fact, become visible."[2] He created diagrams and paintings to help his audiences appreciate and understand the physics of certain cloud formations (as shown on the facing page). And his descriptions of the forms and behavior of clouds are especially vivid: "threads, and meshes, and tresses, and tapestries, flying, failing, melting, reappearing; spinning and unspinning themselves, coiling and uncoiling, winding and unwinding, faster than eye or thought can follow."[3]

Arthur Severn, after John Ruskin, *Cloud Study: Ice Clouds over Coniston,* c. 1884 after 1880 original. Gouache on buff paper, 5 × 6⅝ in. (12.5 × 17 cm). The Ruskin, Lancaster University, 1996P1216.

Ruskin's drawings of the sky are at the same time boldly expressive works of art and records of their maker's close study of natural phenomena. Ruskin's observations of the changes to the English sky resulting from industrial pollution prompted him to prepare a lecture series titled "The Storm-Cloud of the Nineteenth Century," considered one of the earliest published commentaries on human-driven climate change. His drawings were copied by his nephew-in-law, Arthur Severn, and engraved to illustrate his lectures.

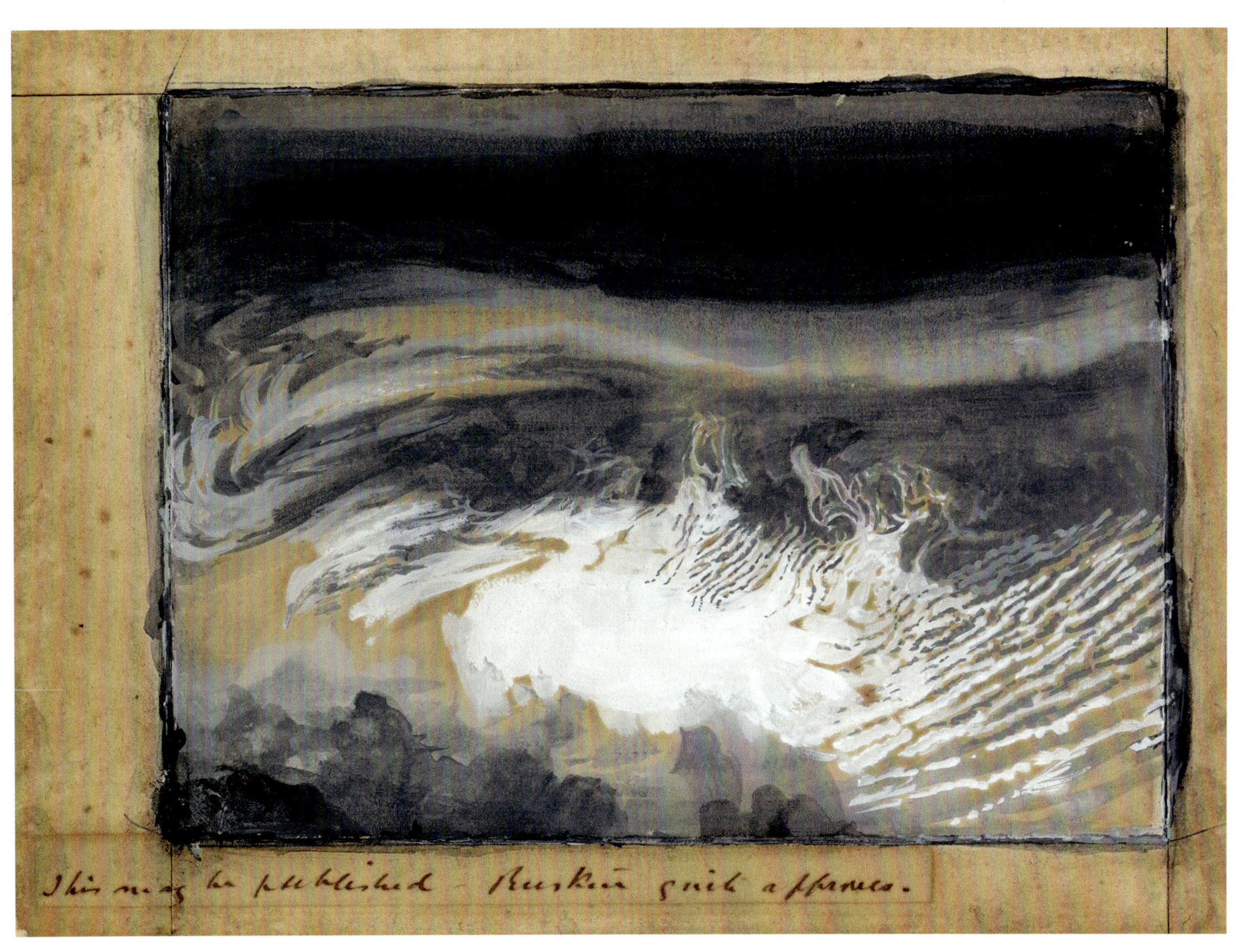
This may be published – Ruskin quite approves.

Ruskin's close and extensive meteorological studies led him to notice changes in weather patterns that he attributed to the effects of industrialization, placing him among the early observers of what we now call anthropogenic climate change.

In *Any-Instant-Whatever* (shown on the facing page), I address issues of climate change and environmental justice, as the sky is shared by humans and nonhuman creatures who, through migration, connect Southern California with the rest of the Pacific Flyway, from the far north to the southernmost reaches of our planet. As we change how our skies look and sound—whether through artificial light and noise, structures, or pollutants—they become a space of added threats. The pollutants we release permeate our skies and affect our human and nonhuman kin, often disproportionately harming those who are already most vulnerable.

Baptiste Morizot, in his book *Ways of Being Alive* (2020), argues that the ecological crisis is existential for the many species that face extinction but is also a crisis of sensibility. And the first symptom of this crisis is expressed in the notion of the "extinction of the experience of nature."[4] Although Morizot's focus is on the "impoverishment of what we can feel, perceive, and understand of living beings and the relations we can weave with them," I broaden that desensitization to how we feel, perceive, and understand natural phenomena and, in *Any-Instant-Whatever*, specifically clouds.[5] The blue sky of our atmosphere contains a concentration of human-made carbon pollution that is undoing the very stability of our existence, and yet we continue to appreciate its sublime beauty. To create *Any-Instant-Whatever*, I documented the sky above Los Angeles from dawn to dusk for several days. This prolonged contemplation of the sky attuned my senses to perceive more precisely the magnificence of our atmosphere, whether the never-ending changes in color and value or the constantly shifting cloud formations.

Rebeca Méndez is an artist, graphic designer, and professor at UCLA. She is the founder and director of the Counterforce Lab research studio.

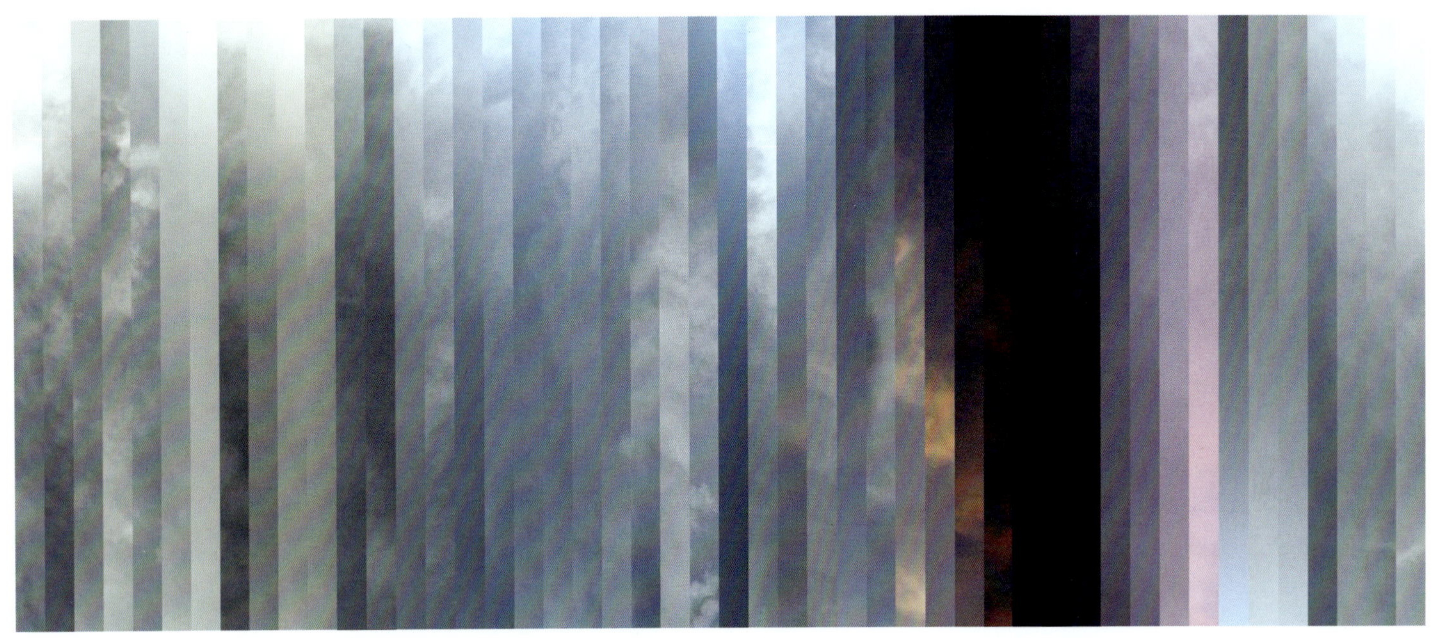

Rebeca Méndez, *Any-Instant-Whatever*, 2020. Film still from 90-minute two-channel looping video. Courtesy of the artist.

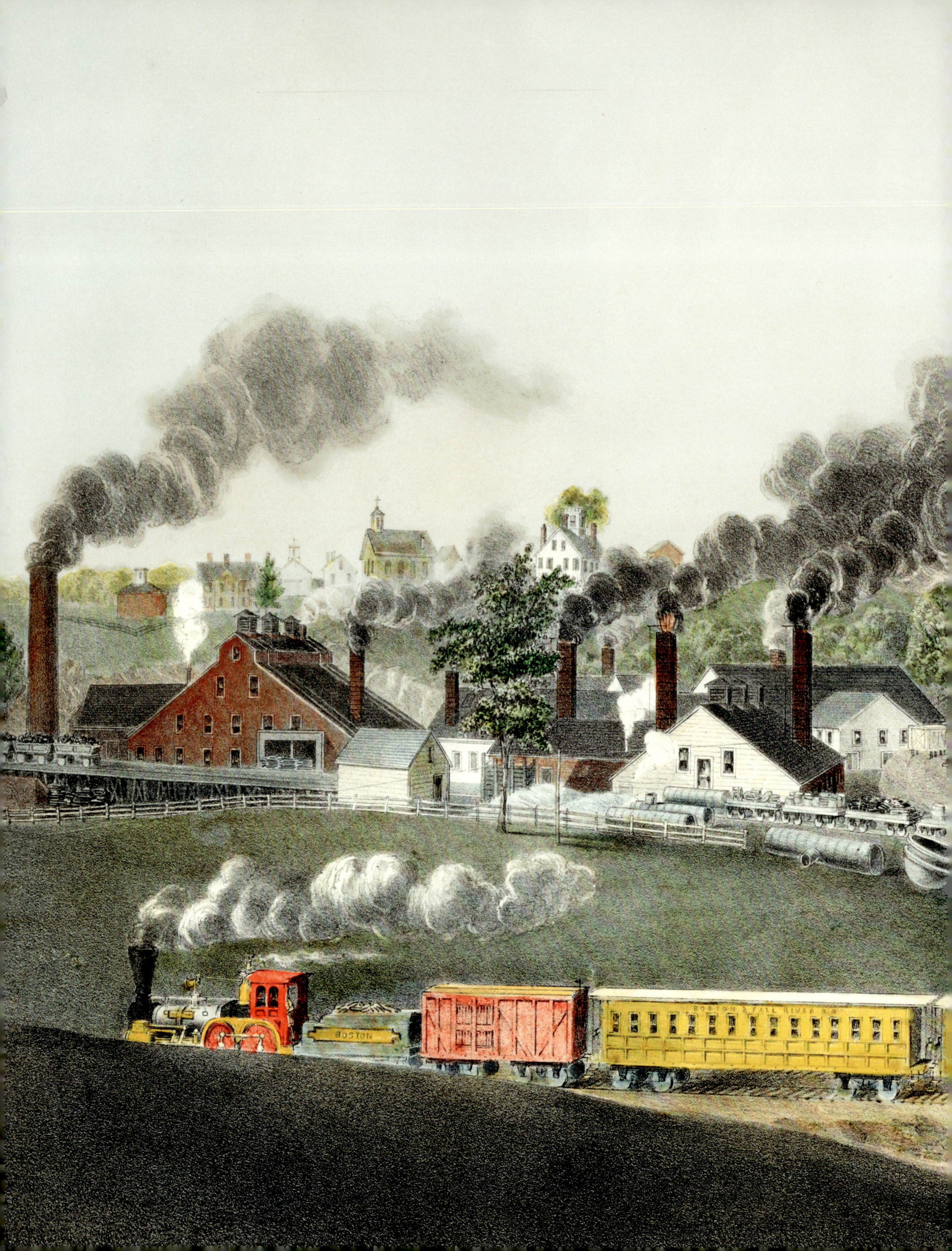
BOSTON
BOSTON & FALL RIVER R.R.

Notes for the Short Essays

Gillian Osborne | Imagining Distorted Scale

1. Edward Hitchcock, *Elementary Geology*, 3rd ed. (1841; New York: Mark H. Newman, 1845), 83–84.
2. Edward Hitchcock, "Geology, Mineralogy, and Scenery of Connecticut," *American Journal of Science and Arts* 7 (1824): 2.
3. Edward Hitchcock, *Reminiscences of Amherst College, Historical, Scientific, Biographical, and Autobiographical* (Northampton, MA: Bridgman & Childs, 1863), 294, 111–12.
4. Robert L. Herbert, "The Sublime Landscapes of Western Massachusetts: Edward Hitchcock's Romantic Naturalism," *Massachusetts Historical Review* 12 (2010): 79.
5. Edward Hitchcock, *The Religion of Geology and Its Connected Sciences* (1851; Boston: Phillips, Sampson, 1859), 24, 413.
6. Hitchcock, *Elementary Geology*, 217.
7. George Perkins Marsh, *Man and Nature; or, Physical Geography as Modified by Human Action* (New York: Charles Scribner, 1864), 13, 36.
8. Hitchcock, *Elementary Geology*, 217–18.

Karen Lloyd | Viewing Stations in the English Lake District

1. Jonathan Bate, *Radical Wordsworth: The Poet Who Changed the World* (London: William Collins, 2020). For "disruptive influence," see Jonathan Bate, "The Romantic Lakes from Wordsworth to Beatrix Potter," transcript of lecture given at Gresham College, London, December 11, 2018, https://www.gresham.ac.uk/watch-now/romantic-leakes-wordsworth-beatrix-potter.

Rachel Storer | Skying

1. Constable's letter to Rev. John Fisher, October 23, 1821, quoted in John E. Thornes, *John Constable's Skies: A Fusion of Art and Science* (Birmingham, UK: University of Birmingham Press, 1999), 199.

Veront M. Satchell | The Hope Lands, Saint Andrew Parish, Jamaica, in 1826

1. Richard Sheridan, "Wealth of Jamaica in the Eighteenth Century," *Economic History Review* 18, no. 2 (August 1965): 292–311.
2. Trevor Burnard, "'Prodigious Riches': The Wealth of Jamaica before the American Revolution," *Economic History Review* 54, no. 3 (2001): 505–13.
3. Climate Change Knowledge Portal for Development Practitioners and Policy Makers, https://climateknowledgeportal.worldbank.org/country/jamaica/climate-data-historical.
4. USAID, "Still Standing, Still Serving in the Midst of Extreme Weather," https://www.usaid.gov/jamaica/news/may-11-2021-still-standing-still-serving-midst-extreme-weather.
5. National Library of Jamaica, "History of Hurricanes and Floods in Jamaica," nlj.gov.jm/history-notes/History%20of%20Hurricanes%20and%20Floods%20in%20Jamaica.pdf.

Suzanne Pierre | There Is No Painting Over a Colonized Ecosystem

1. Kathryn Yusoff, "Geology, Race, and Matter," in *A Billion Black Anthropocenes or None* (Minneapolis: University of Minnesota Press, 2018), https://manifold.umn.edu/projects/a-billion-black-anthropocenes-or-none.

Dennis Carr | Thomas Cole's *Portage Falls on the Genesee*

1. See Samuel B. Ruggles to William H. Seward, New York, July 24, 1841, University of Rochester Library. On the painting, see also William Cullen Bryant II, "Poetry and Painting: A Love Affair Long Ago," *American Quarterly* 22, no. 4 (Winter 1970): 859–82.
2. See Elizabeth Mankin Kornhauser and Tim Barringer, *Thomas Cole's Journey: Atlantic Crossings* (New York: Metropolitan Museum of Art, 2018), 72–74.

Kim Stanley Robinson | John Muir in 2024

1. Sammy Roth, "He Ran the NAACP: Now He's Leading the Sierra Club's Fight against Climate Change," *Los Angeles Times*, February 23, 2023, sec. Climate & Environment; https://www.latimes.com/environment/newsletter/2023-02-23/ben-jealous-led-the-naacp-now-hes-running-the-sierra-club-boiling-point.

Alan Bacock | A *Paya* (Water) Story

1. Owens Valley Groundwater Authority, "The Plan," https://ovga.us/the-plan/.

Kristen Anthony | The Promise of Smoke

1. Helen A. Cooper, ed., *Life, Liberty, and the Pursuit of Happiness: American Art from the Yale University Art Gallery* (New Haven: Yale University Art Gallery in association with Yale University Press, 2008), 267.
2. David Stradling, *Smokestacks and Progressives: Environmentalists, Engineers, and Air Quality in America, 1881–1951* (Baltimore, MD: Johns Hopkins University Press, 1999), 3.
3. Joseph Pennell, "XII Edgar Thomson Steel Works," in *Pictures of the Wonder of Work* (Philadelphia and London: J. B. Lippincott, 1916), n.p.
4. Stradling, *Smokestacks and Progressives*, 22.
5. Greta Thunberg, transcript of speech delivered at the United Nations Climate Action Summit, New York, NY, September 23, 2019, https://www.npr.org/2019/09/23/763452863/transcript-greta-thunbergs-speech-at-the-u-n-climate-action-summit.

Jonathan S. Blake | The Unlovely Forests of Los Angeles

1. Inglewood Oil Field, "FAQ," https://inglewoodoilfield.com/faq/; Clean Break: From Dirty Fossil Fuels to Clean Renewable Energy, "Baldwin Hills Inglewood Oil Field," https://cleanbreak.info/la-county-drilling-baldwin-hills-inglewood-oil-field/.
2. Upton Sinclair, *Oil!* (New York: Albert & Charles Boni, 1927), 109.
3. "Venice Battle Attests Oil and Water Do Mix," *Los Angeles Times*, June 29, 1930.
4. Sinclair, *Oil!*, 526.
5. Ibid., 527.

Rebeca Méndez | The Storm-Cloud of the Twenty-First Century

1. Stephen Nowlin, *Sky*, exh. cat. (Pasadena, CA: Art Center College of Design, 2020), [18].
2. John Ruskin, *The Storm Cloud of the Nineteenth Century* (Orpington, UK: G. Allen, 1884), 19.
3. Ibid., 29.
4. Baptiste Morizot, "Introduction: The Ecological Crisis as a Crisis of Sensibility," in *Ways of Being Alive*, trans. Andrew Brown (Cambridge, UK: Polity Press, 2020), 6.
5. Ibid.

List of Lenders

Amherst College Archives and Special Collections, Amherst, MA
Ashmolean Museum, Oxford
Autry Museum of the American West, Los Angeles
Brantwood, Cumbria, UK
The Sterling and Francine Clark Art Institute, Williamstown, MA
Colonial Williamsburg Foundation, VA
Concord Free Public Library, Concord, MA
Concord Museum, Concord, MA
Copperfield Gallery, London
Delaware Art Museum, Wilmington
FIDM Museum, Los Angeles
William (Ned) Friedman
Erkki Huhtamo Media Archeology Collection
Ethan and Joanne Lipsig
Los Angeles County Museum of Art
Rebeca Méndez
National Gallery of Art, Washington, DC
Natural History Museum, London
Natural History Museum of Los Angeles County
New York Public Library
Royal Meteorological Society, Science Museum Group, London
The Ruskin, Lancaster University, UK
Leah Sobsey
Tate Britain, London
University of California, San Diego, Special Collections and Archives
Wadsworth Atheneum Museum of Art, Hartford, CT
Wellcome Collection, London
Will Wilson
Wordsworth Trust, Grasmere, UK
Yale Center for British Art, New Haven

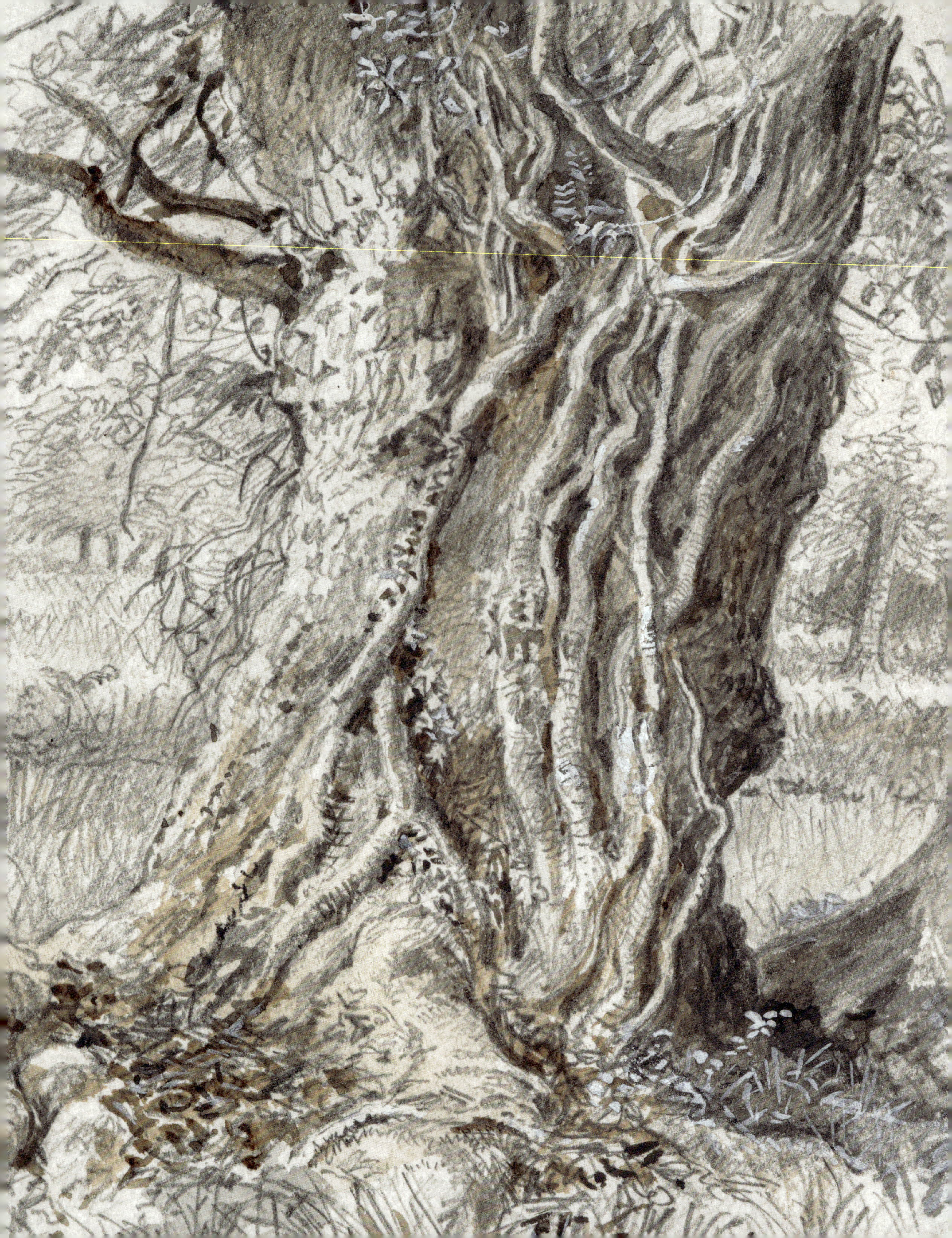

Acknowledgments

A project like *Storm Cloud: Picturing the Origins of Our Climate Crisis*, which includes both a book and an exhibition, could not have happened without the support of a great many people. First and foremost, we want to thank the project's assistant curator, Kristen Anthony, whose research and organizational skills, enthusiasm, intellectual curiosity, and commitment to sustainability made both the exhibition and the book exponentially better. The project's curatorial team had not worked together before and it was a deeply educational experience, a satisfying collaboration, and—amid the logistics and climate doom—often a joy.

Storm Cloud was born in early 2020 as a small exhibition intended to highlight The Huntington's collections. During the wild uncertainties of the pandemic, it grew in both square footage and intellectual purpose, an expansion that began with the first virtual meeting of the project's advisory council. Our thanks go to Melissa Bailes, Timothy Barringer, Kate Flint, Dehn Gilmore, Nicholas Robbins, Marnie Sandweiss, and Tristram Wolff for suggesting so many of the avenues this project now explores. Alex Kidson, Stevie Ruiz, John Styles, Malcolm Warner, and the late Natale A. Zappia also provided valuable intellectual guidance.

Many others have offered advice and assistance along the way, answering questions both broad and narrow. These include the individual lenders and professionals at the many institutions from which we borrowed materials, including Lynda Classen, Amy Concannon, Jeff Cowton, Martina Droth, Anna Ferrari, William (Ned) Friedman, Colin Harrison, Andrea Hart, Austin Hendy, Juliet Hook, Erkki Huhtamo, Howard Hull, Carol Jacobi, Kevin Jones, Sandra Kemp, Leah Lehmbeck, Anne Leonard, Ethan and Joanne Lipsig, Sophie Lynwood, Erin Monroe, Maggie Mustard, Susan Oshima, Nate Smith, and Carolyn Vega.

We are additionally grateful to our Huntington curatorial colleagues Peter Blodgett, Erin Chase, Lauren Cross, Josh Garrett-Davis, Sean Lahmeyer, Linde Lehtinen, Daniel Lewis, David Mihaly, Krystle Satrum, Clay Stalls, Olga Tsapina, and Vanessa Wilkie for their help navigating the institution's vast collections. The exhibition drew widely on the expertise and collections of the Library, Art, and Botanical divisions. A shout-out is also due to *Storm Cloud*'s Cli-Fi Book Club, and especially Sara K. Austin and Michele Navakas, with whom we learned so much while discussing the weird world of Victorian speculative fiction.

The exhibition itself would not have happened without the work of teams across The Huntington. In the Exhibitions Department, Lana Johnson, Angela Fann, and Danielle Killam made sure all the moving parts came together on time. Registrars Katie Leavens and Jenny Werner managed both the checklist and installation, with assistance from Lindsey Hansen. The exhibition comprises a wide array of objects—from paintings and drawings, to rare books and

Walter Crane, *Old Oak in Boldrewood* (detail), 1862. Graphite and watercolor on paper, 5¼ × 5 in. (13.3 × 12.7 cm). The Huntington, 71.61.

manuscripts, to rocks and fossils, including a full-scale cast of an ichthyosaurus skull, and oddities like a piece of Confederate-grown cotton. The Huntington's able staff of preparators beautifully installed each one. We are especially thankful for Andrea Perez-Martinez's contributions. Objects from The Huntington's collection were carefully reviewed and, in some cases, conserved by every single member of The Huntington's top-notch Conservation team. Library stacks staff mobilized to retrieve items and tolerated our flotilla of reserve carts during the exhibition planning process. Everything came together finally in an installation designed by Stephen Saitas, with graphics by Debi van Zyl. Their job was a challenge, given the exhibition's strange mixture of beauty and terror. Together they created a viewing experience that encouraged engagement with individual objects and fostered contemplation of the history they reflect.

An exhibition is far more than its physical manifestation. We are grateful to the staff of The Huntington's Advancement Division, overseen by Randy Shulman, Senior Vice President for Advancement and External Relations, and, notably, Sarah Basile, Corporate and Foundation Relations Senior Director, for securing resources for this project. We are hugely grateful to the Getty Foundation and the PST ART: *Art and Science Collide* initiative for 2024–25 for their support for *Storm Cloud* and for creating educational resources and community around sustainable exhibitions. We produced The Huntington's first Climate Impact Report for this exhibition, and are grateful to PST for their guidance, and to our colleagues for believing in the value of such a project, and committing to the extra work that it required, especially Kristen Anthony, Angela Fann, Lana Johnson, Sean Kennedy, and Cindy Tovar. Thanks are also due to our colleagues in Communications and Education who do so much to amplify the impact of exhibitions through their promotion and programming.

Though not all are reproduced in this publication, the work of five contemporary artists was included in the exhibition. Through objects that create dialogue between the practices of the past and our current climate crisis, Binh Danh, Rebeca Méndez, Jamilah Sabur, Leah Sobsey, and Will Wilson connect the dots between the nineteenth century and today.

Long after the exhibition closes its doors, this book will persist as a partial record of that undertaking. Our sincere thanks go to Kristen Anthony, Alan Bacock, Jonathan S. Blake, Dennis Carr, Kristen Case, Nicole Cavender, M Jackson, Karen Lloyd, Rebeca Méndez, Gillian Osborne, Suzanne Pierre, Richard Primack, Nicholas Robbins, Kim Stanley Robinson, Veront M. Satchell, Rachel Storer, and Jan Zalasiewicz for sharing their perspectives on the material. The book itself came together due to the heroic efforts of The Huntington's Publications Department, especially Jean Patterson and Shirin Sadjadpour, and the team in Digital Collections and Imaging Services. Thanks also to the team at Yale University Press: acquiring editor Amy Canonico, managing editor Alison Hagge, production manager Sarah Henry, proofreader Judy Loeven, indexer Enid Zafran, and designer Jeff Wincapaw, who turned a manuscript comprising eighteen essays into this beautiful, coherent volume.

We are truly grateful to The Huntington's senior leadership for their support of this project, especially Sandra Brooke Gordon, Avery Director of the Library, and Christina Nielsen, Hannah and Russel Kully Director of the Art Museum, as well as Huntington president, Karen Lawrence, and the institution's Trustees, who have made sustainability a vital part of our strategic plan.

Thank you also to our families: Ian and Lucas (who unfortunately got too old over the course of this project's development to really appreciate the fact that there are dinosaur—actually, marine reptile—fossils on our checklist), and Mikey Appuhn (who has an abiding interest in the connection between semis and dinos).

Lastly, we wish to state that The Huntington exists on the ancestral lands of the Gabrielino-Tongva and Kizh Nation peoples who continue to call this region home. The Huntington respectfully acknowledges these Indigenous peoples as the traditional caretakers of this landscape, as the direct descendants of the first peoples. The Huntington recognizes their continued presence and is grateful to have the opportunity to work and learn on this land.

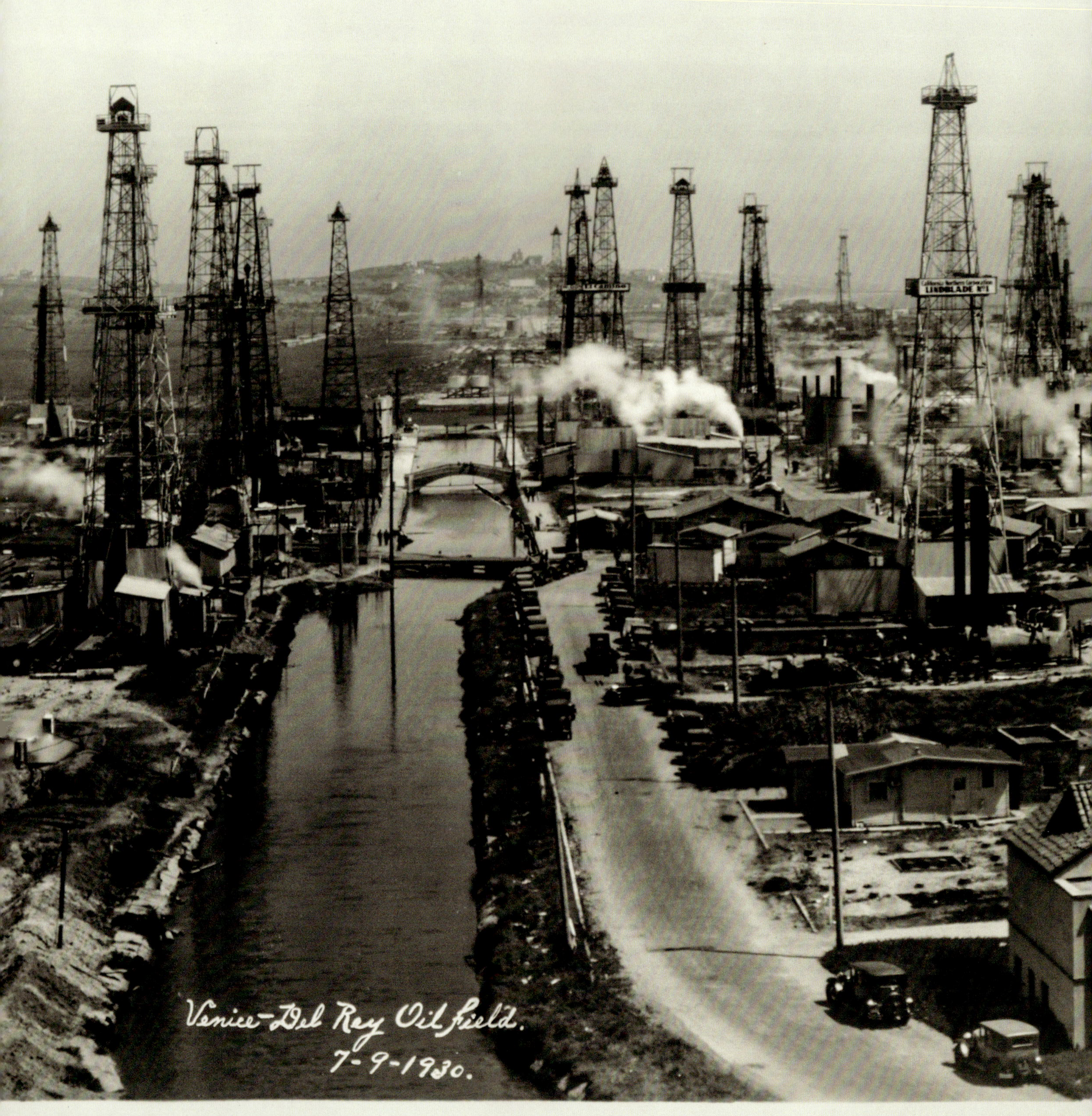
LINDBLADE No. 1
Venice-Del Rey Oil field.
7-9-1930.

Index

Image Credits

© The Ruskin, Lancaster University (fig. 1.1; image on page 149); © CSIRO 2019; Graphic: Scripps Institution of Oceanography at UC San Diego (fig. 1.2); © British Library Board. All Rights Reserved/Bridgeman Images (fig. 1.3); Photos: © Science Museum Group (figs. 1.5, 2.3); © Amgueddfa Cymru—Museum Wales (fig. 1.9); Photo: The University of Chicago Library (fig. 1.13); Photos: Tate (figs. 1.17, 2.4; images on pages 109 and 112); Photo: © Victoria and Albert Museum, London (fig. 2.5); Reproduced with the kind permission of Suffolk Archives and Sudbury Town Council (Sudbury, Suffolk, UK) (figs. 2.6, 2.7); © The Frick Collection, Photo: Joseph Coscia Jr. (fig. 2.8); © Photo: Royal Academy of Arts, London; Photographer: Prudence Cuming Associates Limited (fig. 2.9); Photo: © The Fitzwilliam Museum, Cambridge (fig. 2.11); © Zoe Leonard (fig. 2.23); © Royal Meteorological Society / Science Museum Group (image on page 114); Photo: Allen Phillips/Wadsworth Atheneum (image on page 121); Photo: © Detroit Institute of Arts, USA/Bridgeman Images (image on page 124); Photo: © Christie's Images Limited (images on pages 125 and 126); © Rebeca Méndez (image on page 151).

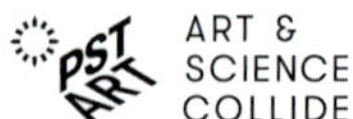

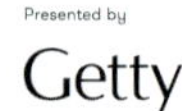

Published on the occasion of the exhibition *Storm Cloud: Picturing the Origins of Our Climate Crisis*, organized by The Huntington Library, Art Museum, and Botanical Gardens, September 14, 2024–January 6, 2025.

This exhibition and its publication are made possible with support from Getty through its PST ART: *Art & Science Collide* initiative.

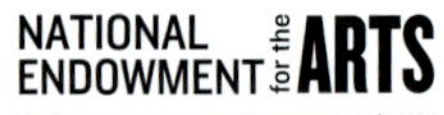

This exhibition and its publication are also supported in part by the National Endowment for the Arts. To find out more about how National Endowment for the Arts grants impact individuals and communities, visit www.arts.gov.

Generous support for this exhibition is provided by the Tianqiao and Chrissy Chen Science Initiative and the Douglas and Eunice Erb Goodan Endowment. Additional funding is provided by The Gladys Krieble Delmas Foundation, The Neilan Foundation, The Ahmanson Foundation Exhibition and Education Endowment, The Melvin R. Seiden-Janine Luke Exhibition Fund in memory of Robert F. Erburu, and the Boone Foundation.

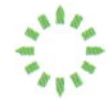

Storm Cloud: Picturing the Origins of Our Climate Crisis participated in the PST ART Climate Impact Program. Learn more at pst.art/climate.

yalebooks.com/art

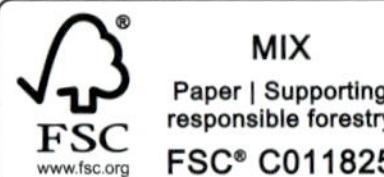

Designed by Jeff Wincapaw
Set in Tisa Pro type
Separations by Embassy Graphics
Printed in Canada by Transcontinental

For The Huntington:
Manager of Book Publishing: Jean Patterson
Publication Coordinator: Shirin Sadjadpour

For Yale University Press:
Editor: Amy Canonico
Assistant Managing Editor: Alison Hagge
Production Manager: Sarah Henry
Editorial Assistant: Elizabeth Searcy

Library of Congress Control Number: 2024937277
ISBN 978-0-300-27614-5

A catalogue record for this book is available from the British Library.

The paper in this book meets the requirements of ANSI/NISO Z39.48–1992 (Permanence of Paper).

10 9 8 7 6 5 4 3 2 1

Cover illustration: (*front*) detail of Arthur Severn, after John Ruskin, *Thunderclouds, Val d'Aosta* (fig. 1.1)
Endpapers: details of "Ideal Section of a Portion of the Earth's Crust" (fig. 1.11)
Title spread: William Pickett's "Iron Works of Coalbrook Dale" (after Philippe Jacques de Loutherbourg) (fig. 1.6)
Contents spread: detail of *Owens Lake* (fig. 1.23)
Essays spread: detail of William Dyce's *Pegwell Bay, Kent* (fig. 1.17)
Short essays spread: detail of Thomas Cole's *Portage Falls on the Genesee* (on page 125)

The extended captions that appear on pages 101–51 were written by Melinda McCurdy, Karla Nielsen, and Kristen Anthony.

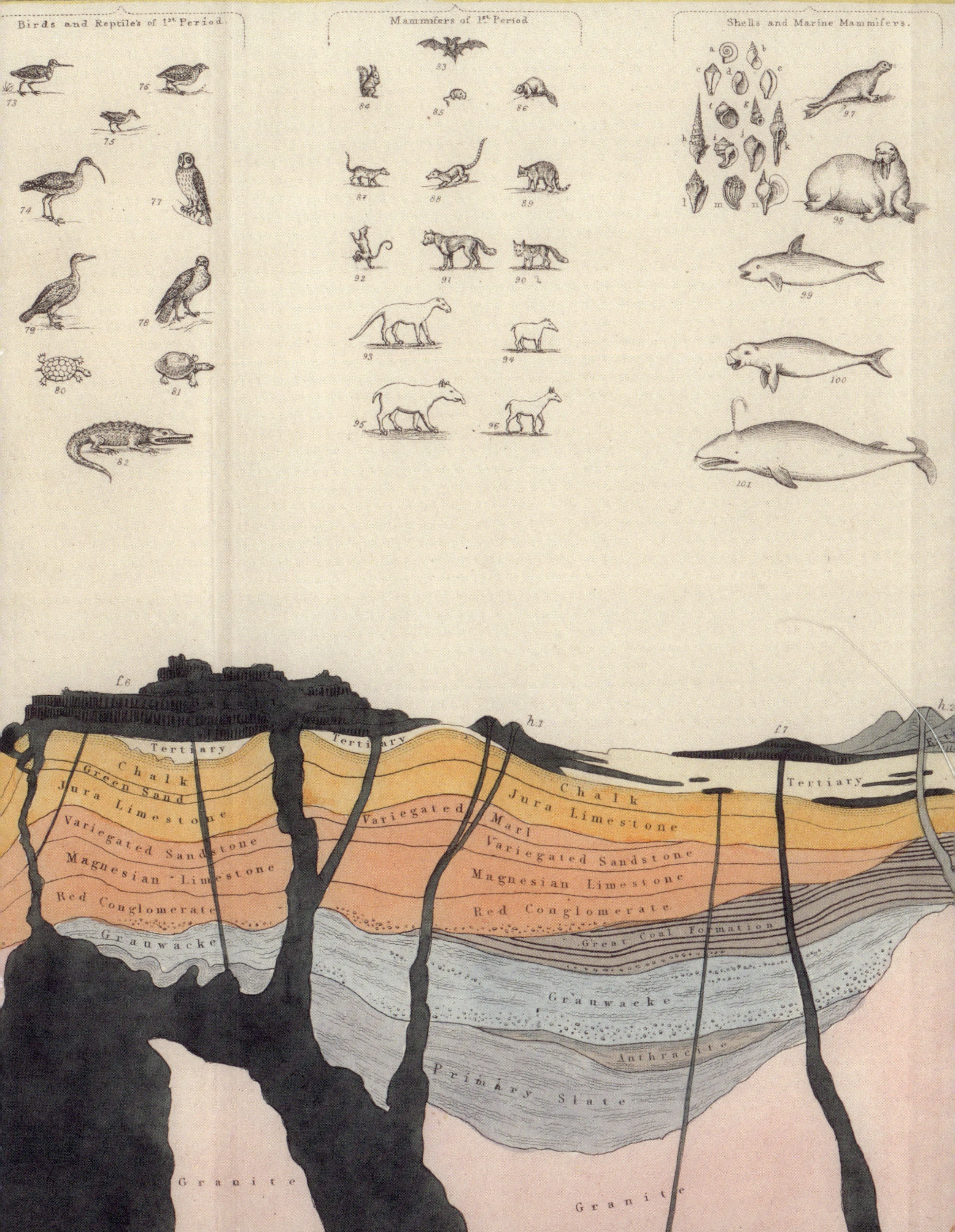
Birds and Reptiles of 1st Period
Mammifers of 1st Period
Shells and Marine Mammifers.
Tertiary
Chalk
Green Sand
Jura Limestone
Variegated Sandstone
Magnesian Limestone
Red Conglomerate
Grauwacke
Variegated Marl
Great Coal Formation
Anthracite
Primary Slate
Granite
f.6
h.1
f.7
h.2

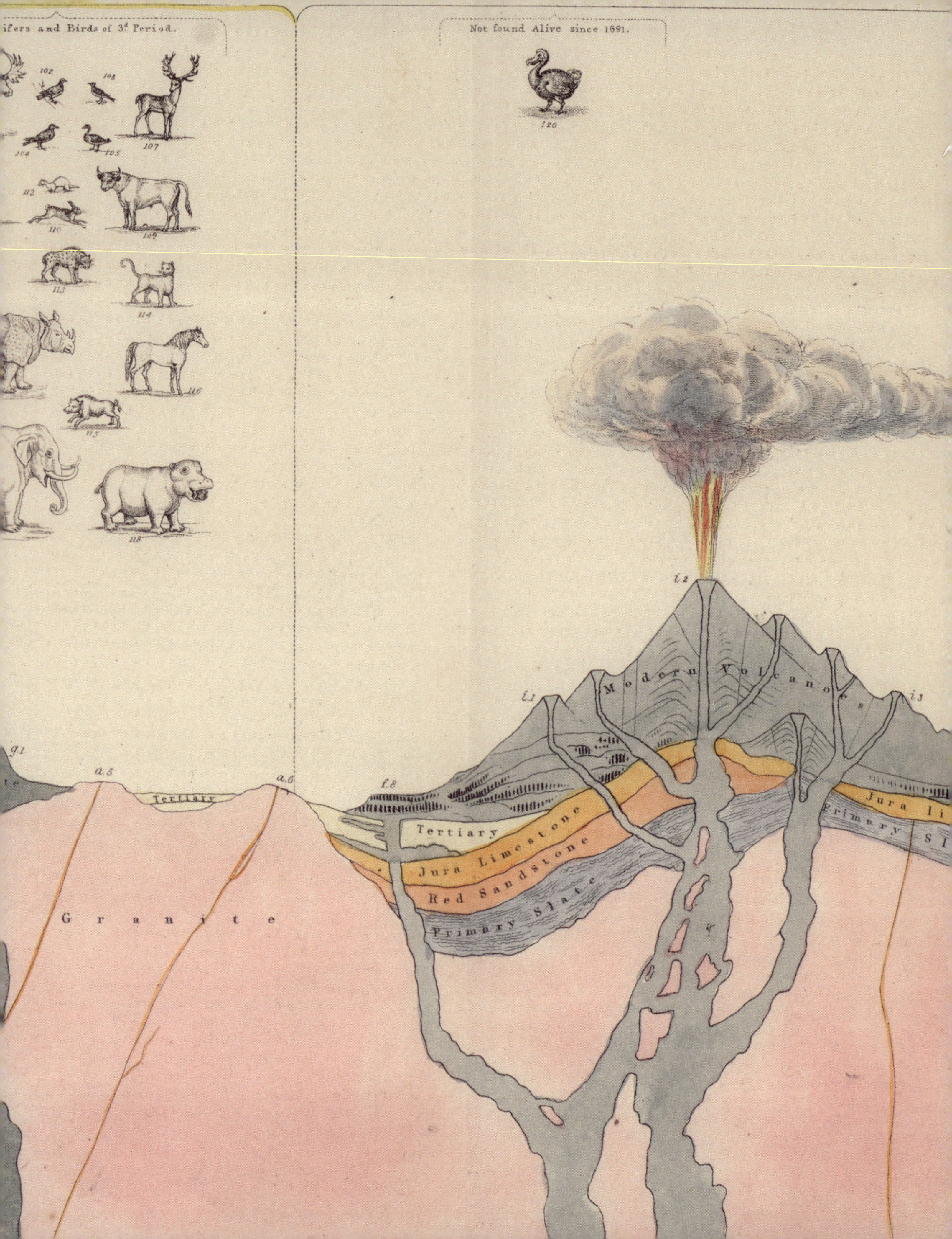
ifers and Birds of 3d Period.
Not found Alive since 1681.
Modern Volcanoes
Tertiary
Jura Limestone
Red Sandstone
Primary Slate
Granite
Jura li
Primary Sl